A mid-Ordovician brachiopod evolutionary hotspot in southern Kazakhstan

by

Leonid E. Popov and L. Robin M. Cocks

Acknowledgements

Financial support for the publication of this issue of Fossils and Strata was provided by the Lethaia Foundation

Contents

A mid-Ordovician brachiopod evolutionary hotspot in southern Kazakhstan

LEONID E. POPOV AND L. ROBIN M. COCKS

FOSSILS AND STRATA

THE LETHAIA FOUNDATION

Popov, L.E. & Cocks, L.R.M. 2021: A mid-Ordovician brachiopod evolutionary hotspot in southern Kazakhstan. *Fossils and Strata*,

Chu-Ili, now in Kazakhstan, was a substantial independent equatorial microcontinental terrane in Ordovician times, with a small Precambrian core fringed by several island arcs. Its mid-Ordovician (late Darriwilian to early Katian) faunas were a major evolutionary hotspot within an equatorial archipelago at a period when Palaeozoic sea levels and temperatures were at their highest. As well as reviewing the previously described brachiopods from elsewhere in Chu-Ili, the mid-Ordovician brachiopods of the West Balkhash Region, which outcrops west of Lake Balkhash within Chu-Ili, are newly described here, mainly from the Berkutsyur and Baigara formations. Many represent the earliest occurrence of their lineages, notably the oldest member of the Order Atrypida. More than twelve brachiopod associations are defined, many for the first time and together hosting 73 genera and over 91 species. The new family Kellerellidae is erected within the superfamily Lissatrypoidea. New genera are *Aploobolus* (Obolidae), *Doughlatomena* (Rafinesquinidae), and *Altynorthis*, *Lictorthis*, and *Baitalorthis* (all Plectorthidae), *Baitalorhynchus* (Sphenotretidae), *Lydirhyncha* (Ancistrorhynchidae) and *Costistriispira* (Kellerellidae). Eleven new species, including *Aploobolus*? *tenuis*, *Doughlatomena splendens*, *Bimuria karatalensis*, *Apatomorpha akbakaiensis*, *Lepidomena betpakdalensis*, *Sonculina baigarensis*, *Altynorthis betpakdalensis*, *Altynorthis vinogradovae*, *Phaceloorthis*? *corrugata*, *Batailorhyncha rectimarginata* and *Costistriispira proavia*, and one new subspecies *Sowerbyella* (*Sowerbyella*) *verecunda baigarensis* are also erected. The global palaeogeographical affinities of all the Chu-Ili brachiopod faunas are discussed, as well as Chu-Ili's place within the peri-Gondwanan archipelago. Newly named stratigraphical units are the Berkutsyur (Darriwilian to early Sandbian) and overlying Kopkurgan (Sandbian to Katian) formations within West Balkhash, and the Tastau (Darriwilian) and Takyrsu (Darriwilian to early Sandbian) formations within the northern Betpak-Dala desert. □ *Brachiopoda, Evolutionary hot spot, Kazakhstan, Ordovician.*

Leonid E. Popov [leonid.popov@museumwales.ac.uk], Department of Geology, National Museum of Wales, Cathays Park Cardiff CF10 3NP, UK; L. Robin M. Cocks [r.cocks@nhm.ac.uk], Department of Earth Sciences, The Natural History Museum, Cromwell Road London SW7 5BD, UK; manuscript received on 23/04/2020; manuscript accepted on 3/07/2020.

Introduction

During the mid-Ordovician, sea levels and temperatures were at their highest in the whole Palaeozoic, and second highest only to the mid-Cretaceous in the entire Phanerozoic, as known from many papers, reviewed by Torsvik & Cocks (2017). In those mid-Ordovician times (late Darriwilian to early Katian), the Chu-Ili Terrane formed part of an archipelago which straddled the Equator, comparably to the East Indies today (Popov & Cocks 2017). Some Middle and Late Ordovician brachiopods from several sectors of the Chu-Ili Terrane have been described, but not from the West Balkhash Region which are thus monographed here. The brachiopods described in previous works from the Chu-Ili Terrane include the Anderken Formation (Sandbian) (Popov *et al.* 2002), the Dulankara Formation (early Katian) (Popov *et al.* 2000;

Popov & Cocks 2006), and the Uzunbulak Formation (Darriwilian) (Nikitina *et al.* 2006). Our aim here is to complete the task for the Middle Ordovician of the entire terrane so as to get a clear picture of its biological and geological significance. We also define more than 12 associations dominated by brachiopods, some previously unrecognised. The freshly described faunas came mostly from the newly defined Berkutsyur and Kopkurgan formations, which outcrop south-west of Lake Balkhash in central Kazakhstan (Fig. 1) as well as from the Baigara Formation together with the new Tastau and Takyrsu formations exposed between the Karatal River and Baigara Mountain in the inhospitable southern Betpak-Dala desert, where access is particularly difficult.

Our conclusion is that Chu-Ili hosted one of the most rapidly-evolving brachiopod faunas in the world in those times, and fully justifies identification as a 'hotspot'. The terrane is thus of cosmopolitan

DOI 10.1002/9781119782377 © 2021 Lethaia Foundation. Published by John Wiley & Sons Ltd

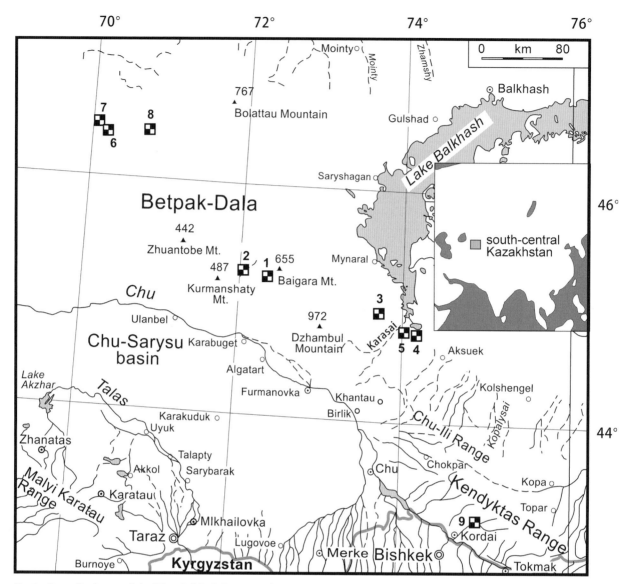

Fig. 1. Generalised map of the West Balkhash Region and eastern Betpak-Dala desert showing the position of the studied areas. Inset, location of south-central Kazakhstan within Asia. Brachiopod localities: 1, area 6 km south west of Baigara Mountain; 2, River Karatal; 3, area 15 km west of Chimpek Bay (Locality 388); 4, area 4 km south-west of Lake Alakol; 5, area 7 km south-west of Lake Alakol; 6, Golubaya Gryada (localities 562, 563); 7, Karakan Ridge (Locality 154); 8, Ergenekty Mountains (Locality 1501); 9, Talapty site.

importance, and played a key role in the Great Ordovician Biodiversification Event (Webby *et al.* 2004). The justification for that conclusion is within the 'Implications for biodiversity' section below.

Geological setting and key sections

The Chu-Ili Terrane was one of about twenty independent microcontinents and island arcs whose remnants are now preserved in the very large country of Kazakhstan, whose Ordovician rocks were reviewed by Nikitin (1972, 1973), as well as in its adjacent countries of Central Asia, notably Kyrgyzstan,

Uzbekistan, and the north-western part of China including Tarim. In Lower Palaeozoic times they were independent terrane units not far from the vast continent of Gondwana to its north-west (Popov & Cocks 2017; Torsvik & Cocks 2017) and are together known as the 'Kazakh terranes'. Chu-Ili itself is made up of a Proterozoic core with 2.8 Ga zircons and an Early to Late Ordovician accretionary wedge which was caused by the progressive nearing of the adjacent Mynaral - South Dzhungaria Terrane to its north. Chu-Ili and North Tien Shan merged in mid-Silurian time (Popov & Cocks 2017), well after the rocks described in this paper were deposited. However, despite differing statements by some authors

(Degtyarev & Ryazantsev 2007; Bazhenov *et al.* 2012; Wilhem *et al.* 2012), it was not until the Early Devonian that the much larger continent of Kazakhstania, which included Chu-Ili and North Tien Shan as well as several other terranes, became an entity.

South Betpak-Dala

The Ordovician stratigraphy of the southern Betpak-Dala desert between Baigara Mountain at the south-east and the Karatal River at the north-east, all situated on the north-western prolongation of the Chu-Ili Terrane (Fig. 1), was summarised by Esenov *et al.* (1971) and Nikitin *et al.* (1980). The area is on the south-eastern (modern orientation) margin of the Chu-Ili terrane facing the Zhalair-Naiman Fault Zone, which is the Silurian suture (Popov *et al.* 2009). The Ordovician succession in the area is, in ascending order: (1) a thick succession of graded siliciclastic rocks with subsidiary units of fine rhyolitic tuffs and a few horizons of phosphoritic conglomerates assigned to the Karatal Formation (Floian to early Darriwilian) with an estimated thickness up to 2000 m and whose stratigraphical relationship with the underlying units remains unknown; (2) the Baigara Formation (late Darriwilian to early Sandbian), a transgressive succession of polymict conglomerates, intercalated brownish-red and greenish-grey sandstones and siltstones with lingulides, succeeded by intercalating units of siltstones, argillites and nodular limestones with abundant dasyclad algae; (3) a succession of graded siliciclastic rocks with occasional graptolites over 1000 m thick; (4) the Anderken Formation (Sandbian) of mainly sandstones and siltstones with a few units of polymict conglomerates and bioclastic limestones, 1000–1500 m thick; and (5) the Dulankara Formation (Katian) of polymict conglomerates, sandstones, siltstones, and limestones up to 1600 m thick (Popov & Cocks 2006). Most of the fossil lists in previous publications are based on preliminary identifications, and the existing monographic record is poor. The Floian to Darriwilian graptolites from the Karatal Formation are known from Tsai (1974, 1976), and conodonts from a single locality in that unit were made known by Tolmacheva (2014). A few brachiopod species from the Baigara Formation were published by Nikiforova & Popov (1981), Popov *et al.* (2001), and Bassett *et al.* (2013), but many more are described in the present paper, while the brachiopod fauna from the Anderken Formation was documented by Popov (1980*a*, 1980*b*, 1985), and Popov *et al.* (2002). The rich brachiopod faunas in the Baigara Formation described here were first discovered by T. B. Rukavishnikova in

1952, and were sampled by D. T. Tsai, L. E. Popov and I. F. Nikitin in 1974.

Area 6 km south-west of Baigara Mountain

This is the type area for the Baigara Formation (Figs 1–3). That unit rests with a slight angular unconformity on the graded siliciclastic rocks of the Karatal Formation and contains Darriwilian graptolites in its upper part (Tsai 1976).

The sedimentary succession in ascending order is as follows:

Unit B1. – Polymict conglomerates succeeded upsection by gritstones and intercalated green and brownish red sandstones and siltstones up to 60 m thick with abundant lingulide fragments including *Ectenoglossa* sp. in the upper part.

Unit B2. – Green, fine-grained sandstones, 50 m.

Unit B3. – Intercalated fine-grained calcareous sandstones, calcareous limestones (rhynchonellide shell beds) and siltstones, 22 m thick, with bivalves, trilobite fragments, and abundant brachiopods of the *Ancistrorhyncha* Association (Locality 1020).

Unit B4. – Green, laminated fine-grained sandstones and siltstones, 6 m.

Unit B5. – Bedded light-grey silty limestones with a 9 m silty sandstone in the lower part with abundant brachiopods of the *Scaphorthis–Strophomena* Association (Localities 1021, 765e).

Unit B6. – Green, laminated siltstones, 4 m, above which is an unexposed interval of about 40 m.

Unit B7. – Greenish-grey slightly argillaceous limestones, 4 m thick, with abundant dasyclad algae and brachiopods of the *Altynorthis* Association, with bryozoans, rare trilobites, and cephalopods (Locality 1022).

Unit B8. – Grey nodular limestones with abundant dasyclad algae and bryozoans, 60 m thick, with brachiopods of the *Altynorthis* Association and cephalopods (Locality 1023).

Unit B9. – Green siltstones with thin layers of limestones and limestone nodules, 67 m.

Unit B10. – Grey and greenish grey graded sandstones and siltstones up to 630 m thick, with rare graptolites including *Climacograptus* sp., *Dicellograptus* sp. and *Leptograptus* sp. (Fig. 2; localities 24 and 26).

In the studied transect the Baigara Formation is overlain conformably by the Anderken Formation.

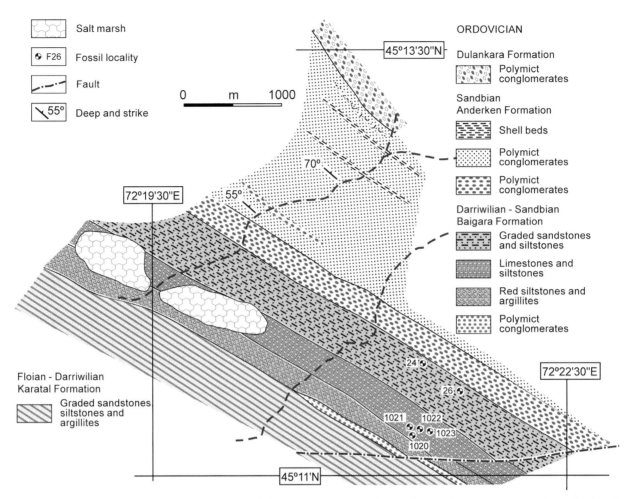

Fig. 2. Schematic geological map of the outcrop area of the Baigara Formation 6 km south-west of Baigara Mountain (Fig. 1, locality 1) showing the positions of the numbered brachiopod localities.

Area on the west side of the Karatal River

This is about 20 km north-west of the Baigara section (Figs 1, 3, 4). D. T. Tsai, L. E. Popov and I. F. Nikitin mapped and measured the section in 1974, which is about 1 km north-east from the River Karatal, where the Baigara Formation rests with minor angular unconformity on graded siliciclastic rocks of the Karatal Formation. The succession in ascending order is as follows:

Unit K1. – Poorly sorted, polymict, brown-grey conglomerates with a sandy matrix, 17 m.
Unit K2. – Intercalations of green and brownish-red sandstones, siltstones, and argillites, with lenses and thin beds of conglomerates in the lower part, 85 m.
Unit K3. – Greenish-grey to brownish-grey siltstones with beds of nodular limestone in the lower part and sandstone in the upper part, total 103 m thick. Limestone beds (Locality 1025) with the *Altynorthis* Association.

Unit K4. – Intercalations of greenish-grey fine-grained sandstones and siltstones with nodules and lenses of a dark-grey, argillaceous limestone, 90 m. Sandstones (Locality 1026) and limestones (Locality 1026b) with abundant *Bimuria-Grammoplecia* brachiopod Association. This unit also occurs on the east side of the River Karatal, at the isolated Locality 1028, 1.4 km south-east from Locality 1026 (Fig. 4).
Unit K5. – Graded greenish-grey and dark-grey sandstones and siltstones up to 1000 m thick. Siltstones in the lowermost part contain the *Bimuria-Grammoplecia* Association (Locality 1026a).

The transition from the Baigara to the Anderken Formation is on the east side of the Karatal River about 2 km west of Sorbulak Spring (Fig. 4, localities 1024, 1024a–c, 1027, and 1027a). Brachiopods of the *Ectenoglossa* and *Tesikella* associations were documented by Popov (1980b, 1985) and Popov *et al.* (2002), and include *Ectenoglossa sorbulakensis, Christiania egregia, Eodalmanella extera, Phragmorthis*

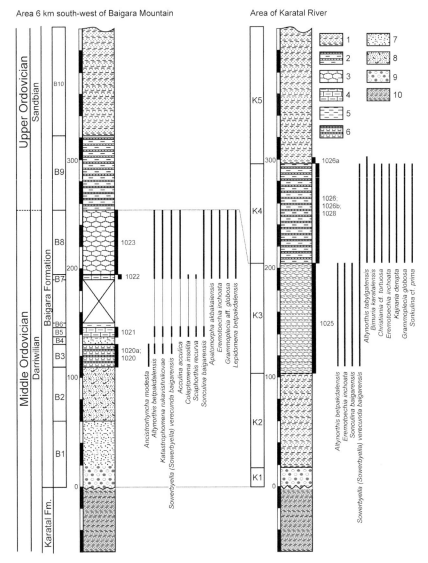

Fig. 3. Schematic stratigraphical sections of the Baigara Formation in the area 6 km south-west of Baigara Mountain and in the area of Karatal River (Fig. 1, localities 1 and 2) showing sample horizons and stratigraphical ranges of selected brachiopod species. Geographical position of the sections and fossil localities are shown on Figs 1, 2, and 4. Lithologies: Baigara Formation (upper Darriwilian – Lower Sandbian): 1, graded sandstones and siltstones; 2, intercalations of siltstones and thin limestone beds; 3, nodular limestones; 4, silty limestones; 5, siltstones; 6, sandstone and siltstone intercalations with brachiopod shell beds; 7, sandstones and siltstones; 8, sandstones with obolid coquinas; 9, polymict conglomerates. Karatal Formation (Floian – Darriwilian): 10, graded sandstones and siltstones.

conciliata, and *Sowerbyella (S.) rukavishnikovae*. The position of the Middle to Upper Ordovician boundary within the Baigara Formation cannot be defined precisely, but *Altynorthis tabylgatensis* and *Sonculina prima* in association with graptolites of the *Nemagraptus gracilis* Biozone in the Tabylgaty Formation in the neighbouring North Tien Shan Terrane (Misius 1986), and *Christiania* cf. *C. tortuosa*, *Grammoplecia globosa*, and *Kajnaria derupta* in units of Sandbian age elsewhere in Kazakhstan, together indicate that Unit K4 in the Karatal section is probably Sandbian.

North Betpak-Dala

Brachiopods from the Middle to Upper Ordovician of North Betpak-Dala are variably known. A rich Darriwilian linguliform brachiopod fauna was described by Nazarov & Popov (1980) from the Karakan Limestone in a barren, unpopulated area informally called the Karakan Ridge (Locality 7 on Fig. 1). Nikitina (1989) described a rhynchonelliform brachiopod fauna from the same locality in her unpublished Ph.D. thesis; and the early Katian rhynchonelliform brachiopods of the *Kellerella* Association from the mud-mound at Sartan-Manai were

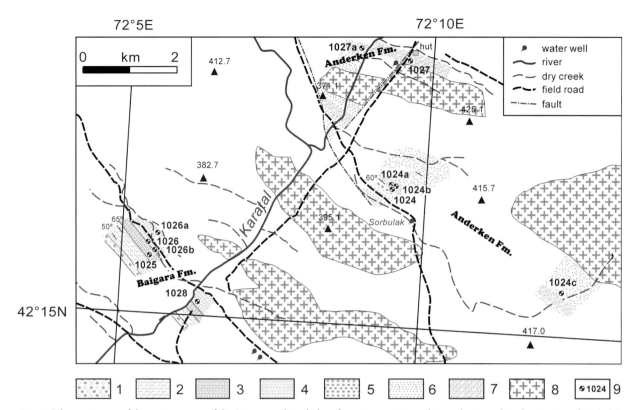

Fig. 4. Schematic map of the outcrop area of the Baigara and Anderken formations at Karatal River basin and in the vicinity of Sorbulak well, south Betpak-Dala (Fig. 1, locality 2) showing positions of brachiopod localities. Baigara Formation (upper Darriwilian – lower Sandbian): 1, polymict conglomerates; 2, green and brownish red sandstones, siltstones, and argillites; 3, nodular limestone and siltstone intercalations; 4, graded sandstones and siltstones. Anderken Formation (Sandbian): 5, polymict conglomerates; 6, sandstones; 7, green and brownish siltstones with argillite intercalations; 8, granitic intrusives.

documented by Nikitin & Popov (1996) and Nikitin *et al.* (1996).

The complex geology of the area was discussed by Keller & Lisogor (1954), Nikitin (1972), Nikitin *et al.* (1980), Nikitina *et al.* (2006), and Ghobadi Pour *et al.* (2009), but is still not yet fully understood. The area was initially surveyed by L. E. Popov in 1973 and 1974 and again between 1986 and 2004 with the late E. V. Alperovich and E. A. Vinogradova. The lithostratigraphy (Kushaky, Savid, Algabas, and Kuyandy formations) previously applied to the Ordovician in the region was imported from the distant Sarysu-Teniz Region by Nikitin *et al.* (1980), Nikitin (1991), and Nikitina *et al.* (2008), but that region is within the North Tien Shan Microcontinent on the opposite side of a Silurian suture (Popov *et al.* 2009; Popov & Cocks 2017) and thus its terminology cannot sensibly be used in the Chu-Ili Terrane. Vinogradova (*in* Ghobadi Pour *et al.* 2009) found that andesite-basalt volcanic rocks assigned to the Savid Formation in the Golubaya Gryada area (Locality 6 on Fig. 1) represent a chain of exhumed intrusive subvolcanic bodies of younger Palaeozoic age which have no relation to the exposed Upper

Ordovician stratigraphical succession as was suggested in some previous publications.

Thus the only existing Ordovician lithostratigraphical name which can correctly be applied in the Betpak-Dala area is the Karakan Formation, whose type area was described by Nikitina *et al.* (2008) and which is a sedimentary succession of bioclastic limestones rich in brachiopods and trilobites, microbial build-ups, and oligomict sandstones, all underlain, overlain, and intercalating with laminated siltstones, black siliceous argillites and cherts with Darriwilian graptolites of the isograptid biofacies. Paragenetic association of the carbonate lithofacies deposited at the environments of inner to mid shelf with distal turbidites and hemipelagites deposited on the continental slope and representing lithofacies of the outer fan and fan fringe is an improbable scenario. We therefore conclude that the carbonate bodies (often referred as the Karakan Limestone) are allochthonous and represent parts of a large slump complex transported by turbidites down the continental slope. The presumed basal units of the Karakan Formation, which include ophiolitoclastic breccias and sandstones, are parts of a serpentinite melange, which includes limestones. Shear lenses of serpentinites and ophiolitic breccias also occur in the limestones and are seen in cores (Nikitina *et al.* 2008), all suggesting that the emplacement of ultramafic volcanic rocks and serpentinites occurred after the Mid Ordovician. Thus, the limestones and fine siliciclastic matrix used by Nikitina *et al.* (2008) as evidence for a Darriwilian age of these heterogenous units must be ignored. Lithologically similar carbonates with comparable faunas preserved as olistoliths also occur in the upper part of the Darriwilian Uzunbulak Formation at Kopalysai, southern Chu-Ili (Nikitina *et al.* 2006). The original sources of these carbonates in both cases might possibly be a collapsed carbonate platform, but such deposits are unknown *in situ* within the region.

Since the lithostratigraphical units designated for the Ordovician of the Baikanur Region are inapplicable, they are replaced here by the new Tastau Formation (Floian to Darriwilian) and Takyrsu Formation (Darriwilian to Sandbian). A further problem is the proper characterisation of the Upper Ordovician (Katian) deposits assigned to the Kuyandy Formation by Nikitin *et al.* (1980) and Nikitin (1991), which remain inadequately known. We conclude that the term 'Karakan Limestone' should be confined to the allochthonous carbonate unit preserved as olistoliths and olistoplacks within the upper part of the Tastau Formation. Yet another dilemma is that the Upper Ordovician Kuyandy Formation of the Baikanur Region is homonymous with the later-named Cambrian (Furongian) Kuyandy Formation of the Boshchekul Region, and the latter name should therefore be replaced in due course.

There is another unnamed upper Ordovician formation exposed on the opposite side of the Chu-Ili Terrane in the Ergenekty Mountains (Fig. 1; site 8) which is 600 m of intercalating sandstones and siltstones with a faulted lower boundary. The upper boundary of the unit is disconformable, with an approximately 350 m unit of brown polymict conglomerates with subsidiary beds of sandstones and siltstones, occasionally with dense raindrop impressions on bedding surfaces in the upper part. The late Ordovician (early to mid Katian) age of the lower unit is confirmed by the brachiopods at Locality 1501 near the Ergibulak well which include *Bokotorthis kasachstanica*, *Qilianotryma* cf. *Q. suspectum*, *Shlyginia* cf. *S. extraordinaria*, and *Sowerbyella (S.) ampla* which are all common in the Otar and Degeres beds of the Dulankara Formation in the southern Chu-Ili Range (Popov *et al.* 1999, 2000; Popov & Cocks 2006).

Tastau Formation (New)

Derivation of name. – The name is derived from Tastau Hill in the type area.

Stratotype. – The type section is the natural exposure cropping out along the ridge informally named by Keller & Lisogor (1954) as Golubaya Gryada (Blue Ridge), north-east of Tastau Hill at 46°26′18″N, 70°23′3″E, altitude 451 m (Fig. 2). It is a remote area in the northern part of the Betpak-Dala Desert (Fig. 1, Locality 6) about 250 km west of Lake Balkhash and 48 km north-east of the Betpak-Dala weather station, which is the closest settlement.

Definition of boundaries. – The lower boundary of the Tastau Formation is poorly defined, but is apparently faulted against serpentinites associated with unspecified Lower Palaeozoic mafic and ultramafic rocks. Keller & Lisogor (1954) and Nikitin *et al.* (1980) reported a thick unit below, the 'Kusheky' Formation of intercalating quartzose and polymict sandstones and siltstones with graptolites of the *Paratetragraptus approximatus* Biozone, including the eponymous species in its uppermost part. Those beds are assigned here to the Lower Member of the Tastau Formation. The upper boundary of the Tastau Formation is faulted in the type area, and previously reported stratigraphical contacts with a so-called 'Savid' Formation are not confirmed by our field observations. However, 15 km south-east, a preliminary visit appeared to show that the Tastau Formation has conformable stratigraphical contact with the late Darriwilian to Sandbian Takyrsu Formation, but that requires confirmation.

Description. – At the type area the Ordovician rocks form a steep monocline dipping at 80° north-east. The Tastau Formation can be divided into three members in ascending order: the Lower Member, mainly graded, light grey, arkosic and quartzose sandstones and subsidiary siltstones, with thickness over 500 m; the Middle Member of bluish-grey to black, graded fine-grained arkosic and quartzose sandstones, siltstones, laminated siliceous siltstones, argillites, and radiolarian cherts, up to 320 m thick; and the Upper Member, which in the type area consists of *c.* 120 m grey to brownish-red argillites and up to 50 m of yellowish grey siliceous argillite and siltstone intercalations. Further west, the Upper Member includes in its lower part an olistostrome complex with large olistoliths and olistoplacks of shallow marine carbonates known as the Karakan Limestone whose thickness increases up to 250 m

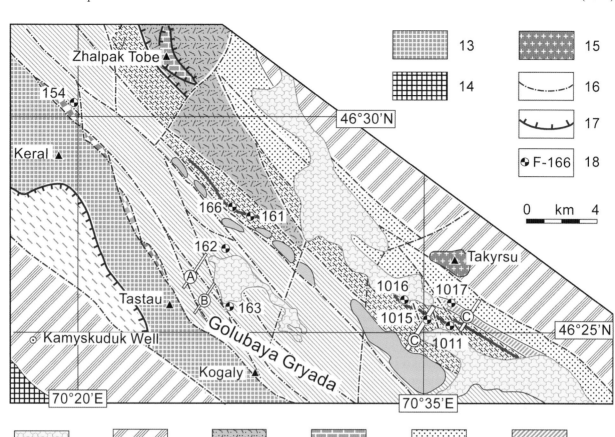

Fig. 5. Schematic geological map of north Betpak-Dala (Fig. 1, localities 6 and 7), showing position of the type sections of newly designated lithostratigraphical units and fossil localities (modified from Ghobadi Pour *et al.*, 2009). Lithologies: 1, salt marshes; 2, Devonian (unspecified); 3, Silurian? volcanic rocks and tuffs; 4, Upper Ordovician limestones (unspecified); 5–8, Takyrsu Formation (late Darriwilian to Sandbian): 5, intercalating sandstones and siltstones, horizons of limestones and conglomerates; 6, graded sandstones and siltstones; 7, intercalating sandstones and siltstones, horizons of tuffs, limestones and conglomerates; 8, marker limestone beds; 9, Karakan Limestone (Darriwilian olistostrome complex); 10, Tastau Formation of graded siltstones, sandstones, siliceous argillites and radiolarian cherts; 11–13, unspecified Lower Palaeozoic rocks: 11, siliciclastic rocks; 12, subvolcanic diabase intrusions; 13, ultramafic rocks and serpentinites; 14, Precambrian; 15, Devonian granites; 16, faults; 17, thrust faults; 18, fossil localities. A-B, type area of the Tastau Formation: A, position of section described by Dubinina *et al.* (1996); B, position of section described by Nikitina *et al.* (2008) and Tolmacheva (2014). C, type section of the Takyrsu Formation.

(Dubinina *et al.* 1996; Nikitina *et al.* 2006; Tolmacheva 2014).

The Tastau Formation crops out more or less continuously in the northern Betpak-Dala desert for about 50 km north-east of the Zhalair Naiman Fault Zone, south-east of the Zhidely dry river. Further south-east it is apparently replaced by the Karatal Formation, which is mainly graded sandstones and siltstones with subsidiary thin layers of tuffs. By comparison with the latter unit, the volcanic tuffs in the Tastau Formation are relatively minor, while the proportion of radiolarian cherts and siliceous argillites (representing background deposits) is considerably greater. The Middle Member contains graptolites characteristic of the *Pendeograptus fruticosus*, *Didymograptus protobifidus*, and *Isograptus maximodivergens* biozones, and conodonts of the *Oepikodus evae* and local *Periodon flabellum/Periodon macrodentatus* biozones (Dubinina *et al.* 1996; Tolmacheva 2014), indicating a late Floian to early Darriwilian age. The Upper Member contains conodonts of the local *Paroistodus horridus* Biozone in its lowermost part (Nikitina *et al.* 2008) and graptolites of the local *Paraglossograptus tentaculatus* and *Pseudoclimacograptus romanovskyi* biozones, indicating a Darriwilian age. Other fossils reported from the Upper Member include the trilobite *Pricyclopyge* sp. and the brachiopods *Akadyria simplex*, *Broeggeria* cf.

B. putilla, and *Oxolosia*? sp. (Fig. 5; localities 154, 162, and 163). Two earlier brachiopod species are also known from the lower part of the Uzunbulak Formation at Kurzhaksai (Nikitina *et al.* 2008; localities 106 and 111a), where they co-occur with graptolites of the *Paraglossograptus tentaculatus* Biozone.

The Tastau Formation was deposited in the lower part of the continental slope facing the Zhalair-Naiman back-arc basin and shows a general retrogradational trend varying from mid-fan (Lower Member) to fan fringe (Upper Member). Nikitina *et al.* (2008) and Tolmacheva (2014) recognised five parasequences within the Tastau Formation Middle Member varying in thickness from 40 to 110 m. The three lower parasequences show an upward-retrogradational trend which ended with the deposition of finely laminated argillites and radiolarian cherts, while the stacking patterns of the two uppermost sequences vary in gradation. The abundance of radiolarian cherts in pelagic sediments and the occurrence of the isograptid graptolite biofacies indicates that sedimentation of the Tastau Formation occurred on the continental margin facing the major marine basin (Popov *et al.* 2009; Popov & Cocks 2017), since neither isograptid biofacies, nor radiolarian oozes would be 'sandwiched' within shallow marine carbonate succession (e.g. the Karakan Limestone) except in exceptional circumstances (Fortey & Cocks 2003).

Takyrsu Formation (New)

Derivation of name. – The name is derived from Takyrsu Hill (at 46°26′13″N, 70°36′40″E, altitude 449 m) north of the type area.

Stratotype. – The stratotype is the natural exposure south of Takyrsu Hill (Figs 1, 5) in the northern part of the Betpak-Dala Desert about 240 km west of Lake Balkhash and 50–55 km north-east of the Betpak-Dala weather station. Ordovician rocks in the type area are exposed in a monocline deepening northeast at under 30–45° at the west and 15–20° in the east of the outcrop area (Nikitin *et al.* 1980; Ghobadi Pour *et al.* 2009, pp. 328–330, fig. 1, 2).

Definition of boundaries. – Precise definition of the lower and upper boundaries of the unit is difficult due to poor exposure and strong tectonic dislocations. Its lower boundary is probably conformable with the Tastau Formation; however, its nature is complicated by the extensive presence of massive subvolcanic bodies presumably of Devonian age, which were previously erroneously included in the 'Savid' Formation together with the beds of

volcanomict pebbly conglomerate and tuff (e.g. Nikitin *et al.* 1980). The latter are here recognised as the basal unit of the Tastau Formation. The formation is considered as faulted against unspecified (perhaps Silurian) volcanic rocks and tuffs (Fig. 5), but future geological re-mapping of the area may well indicate that it is unconformable in places.

Description. – The Takyrsu Formation comprises (in ascending order): (1) fine to medium grained, often cross-laminated sandstones with a bed of volcanomict pebbly conglomerate and several units of tuffs and tuffaceous sandstones in the lower part and lingulide shell beds (Fig. 5, Locality 1015), up to 300 m thick; (2) bioclastic limestone with 40–50 cm beds of pebbly conglomerate at the base and 1.5 m at the top; (3) grey siliceous argillites with graptolites (Fig. 5, Locality 1016), *c.* 20 m; (4) graded sandstones and siltstones over 350 m thick with graptolites (Fig. 5, Locality 1017). Further information on the sedimentary rocks and units here assigned to the Takyrsu Formation are also in Nikitin *et al.* (1980) and Ghobadi Pour *et al.* (2009).

Distribution. – The Takyrsu Formation is developed in the northern Betpak-Dala area, where it is traceable for over 50 km from the Aschisu dry river to the area south of Takyrsu Hill (Figs 1, 5). The Takyrsu Formation shows a distinct retrogradational trend in its lower part. Unit 1, characterised by polymict conglomerates, cross-laminated sandstones, and the occurrence of large lingulide shells in the upper part, was deposited in a shallow marine environment of the shoreface zone. Occasional beds of andesite tuffs suggest proximity to an active volcanic arc at the time of their formation. Limestones of Unit 2 contain a trilobite assemblage (including *Alperillaenus intermedius*, *Damiraspis margiana*, *Eorobergia* sp., *Farasaphus singularis*, and *Pliomerina* aff. *P. sulcifrons*) together characteristic of the asaphid biofacies confined to inshore environments (Ghobadi Pour *et al.* 2009; 2011). Pebbles from the conglomerate beds include black cherts, quartz, fine siliciclastic, and volcanic rocks, suggesting a growing accretionary wedge as a possible source.

Age. – The late Darriwilian to Sandbian age of the Takyrsu Formation is mainly based on the occasional occurrences of graptolites (Keller & Lisogor 1954; Tsai *in* Nikitin *et al.* 1980). The grey, siliceous, graptolite-bearing argillites at the base of Unit 3 correspond with the maximum flooding surface and are followed by a progradational system of about 370 m of siliciclastic sediments. The late Darriwilian age of the lower part of the Takyrsu Formation is indicated by the graptolites *Hustedograptus teretiusculus*, *Climacograptus* sp., and *Glossograptus* sp. (Fig. 5, Locality 1015) in Unit 3; however, the sporadic occurrence of *Climacograptus* may not exclude a possible early Sandbian age. *Dicranograptus* cf. *D. brevicaulis* and *Pseudoclimacograptus* sp. in Unit 4 (Fig. 5, Locality 1017) also support the late Darriwilian (*Hustedograptus teretiusculus* Biozone) to early Sandbian (*Nemagraptus gracilis* Biozone) age for the Takyrsu Formation.

The lateral lithofacies changes within the Takyrsu Formation remain poorly understood due to inadequate exposure and lack of detailed sedimentological studies. In particular, 17 km northwest of the type section and north of Golubaya Gryada, the succession within the formation appears reversed, with a unit of graptolite-bearing dark grey, graded, fine grained sandstones: argillites at the base and light grey, bedded bioclastic limestones (the 'Kipchak Limestone' of Keller & Lisogor 1954) on the top (Fig. 5, localities 161, 166). The siliciclastic unit (Locality 161) contains graptolites including *Hustedograptus teretiusculus*, *Xiphograptus robustus*, and *Climacograptus*? sp. (Keller & Lisogor 1954; Nikitin *et al.* 1980; Ghobadi Pour *et al.* 2009). The occurrence of *Xiphograptus* strongly indicates a late Darriwilian age for the siliciclastic unit (Maletz 2010).

The overlying Kipchak Limestone (Locality 166) contains the brachiopods *Altynorthis tabylgatensis*, *Buminomena* sp., *Eridorthis* sp., *Esilia* cf. *E. tchetverikovae*, *Ptychoglyptus* sp., *Shlyginia* sp., and *Sowerbyella (S.) verecunda baigarensis*, in addition to an asaphid-dominated trilobite association. That assemblage is similar to those in the limestone bed in the upper part of the Berkutsyur Formation in the West Balkhash Region and is probably early Sandbian, but that conclusion is provisional without more conodonts.

The West Balkhash Region

The Ordovician geology of the West Balkhash Region, including the Sarytuma Fault Zone, Burultas and Mynaral – South Dzhungaria tectonofacies belts, as presented in the publications of Esenov *et al.* (1971) and Nikitin *et al.* (1980), is very outdated. Although significant improvement in understanding the complex Lower Palaeozoic geology was achieved between Nikitin and Popov 1985 and 1993 during extensive geological mapping of the area, under the supervision of the late E. V. Alperovich and E. A. Vinogradova, which clearly demonstrated that the area represents an Early

Palaeozoic accretionary wedge and forearc which subsequently evolved into a foreland basin sealed in the late Katian, as summarised by Popov *et al.* (2009); however, much of the information remains unpublished.

The area south-west of Lake Alakol

The best Middle to Upper Ordovician succession in the West Balkhash Region is seen south-west of Lake Alakol, west of the road connecting Almaty with the town of Balkhash on the northern side of Lake Balkhash (Fig. 1), as outlined by Nikitina *et al.* (2006) and Popov *et al.* (2009). The former publication also contains an account of the Dapingian to Darriwilian brachiopod faunas, but the lithostratigraphy of the area is outdated. The Sandbian deposits in West Balkhash had been assigned to the Oisaksaul and Anderken formations, while the Katian deposits were assigned to the Dulankara, Kyzylsai, and Chokpar formations (Nikitin *et al.* 1980; Nikitin 1991). However, application of the lithostratigraphical subdivision based on the Ordovician succession of the Chu-Ili Terrane core (Zhalair-Naiman Domain) to the contemporaneous deposits of the West Balkhash Region within the Sarytuma Fault Zone is not appropriate due to the different tectonic setting and geological history (Popov *et al.* 2009; Popov & Cocks 2017). The Oisaksaul Formation is also an invalid name, since its type section and boundaries were not made clear. The Middle to Lower Ordovician lithostratigraphy is revised here and includes the new Darriwilian to early Sandbian Berkutsyur Formation and the late Sandbian to middle Katian Kopkurgan Formation. Both units are exposed on the west side of the Lake Balkhash within the Sarytuma Fault Zone between the west side of Lake Alakol and the Burultas valley.

Berkutsyur Formation (New)

Derivation of name. – The name is derived from Berkutsyur Hill about 8 km south-west of the type area.

Stratotype. – The stratotype lies in a natural exposure within the quadrangle of 44°45′ to 44°50′N and 74° to 74°07′30″E (Fig. 1), about 4 km south-west of Lake Alakol. It is exposed in the northern and western wings and the core of a syncline (Fig. 6), but its southern wing is cut by a fault. Thus the contact of the formation with overlying units can best be seen in another and separate block exposed north-east of the main transect.

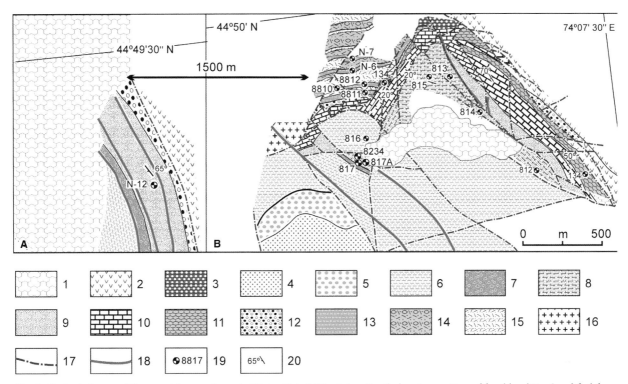

Fig. 6. Geological map of the area 4 km south-west of Lake Alakol (Fig. 1, locality 4) showing position of fossil localities (modified from Nikitina *et al.* 2006 and Popov *et al.* 2009). Lithologies: 1, salt marshes; 2–3, Koktas Formation (Devonian): 2, volcanic rocks and tuffs; 2, conglomerates and sandstones; 4–5, Upper Ordovician (Katian) unspecified rocks; 4, cross- and horizontal-laminated sandstones; 5, polymict conglomerates; 6–7, Kopkurgan Formation (late Sandbian to early Katian): 6, intercalating sandstones and siltstones; 7, olistostrome horizons; 8–12, Berkutsyur Formation (Darriwilian to early Sandbian): 8, siltstones, argillites and argillaceous limestones; 9, arkosic sandstones; 10, limestones; 11, argillaceous and nodular limestones; 12, polymict conglomerates. 13–15, Balgozha Formation (Tremadocian to Dapingian): 13, black siliceous argillites; 14, rhyolite – dacite volcanic mass flow deposits; 15, rhyolite – dacite tuffs; 16, Middle Ordovician intrusive rocks of the Akzhal Complex; 17, faults; 18, dykes; 19, fossil localities; 20, dip and strike.

Definition of boundaries. – The Berkutsyur Formation rests with a minor unconformity on the black siliceous shales, tuffs, and volcanic mass flows of the Tremadocian to Dapingian Balgozha Formation (Nikitina *et al.* 2006). The upper boundary is placed below the unit of black graptolitic shales or olistostrome horizon at the base of the succeeding Kopkurgan Formation.

Description. – In the type section the Berkutsyur Formation is subdivided into eight units in ascending order as follows (Fig. 7):

Unit Be1. – Polymict pebbly conglomerates with sandy matrix, 30 m.
Unit Be2. – Calcareous polymict sandstones 6 m thick with *Aporthophyla kasakhstanica*, *Leptella sarytumensis*, *Martellia reliqua*, *Pomatotrema fecunda*, *Trematorthis karasaiensis*, and *Tuvinia alakulica* (Locality 134).
Unit Be3. – Nodular, sandy and silty limestones, 25 m thick with stromatolites, stromatoporoids, abundant brachiopods, ostracods, and

trilobites (Nikitina *et al.* 2006, localities 133 and 142).
Unit Be4. – Light-grey thick-bedded and massive limestones up to 110 m.
Unit Be5. – Brownish-red and greyish-green, coarse-grained, laminated sandstones up to 120 m.
Unit Be6. – Greyish- and brownish-yellow, coarse-grained, laminated oligomict and arkosic sandstones, with lenses (up to 0.5 m) and a few beds (up to 1.5 m) of polymict conglomerate with well-washed, rounded pebbles of quartz and andesite up to 3 cm across; up to 87 m. In the described transect this unit is barren; however, 2.3 km west at Locality N-6 (Fig. 6) there are beds with disarticulated valves of *Ancistrorhyncha modesta* and unidentified bivalve molluscs.
Unit Be7. – Intercalations of brownish-green to greyish-green coarse-, medium- and fine-grained tuffaceous sandstones, 22 m.
Unit Be8. – Intercalations of brownish yellow and greyish-green fine-grained calcareous sandstones and siltstones with estimated thickness up to

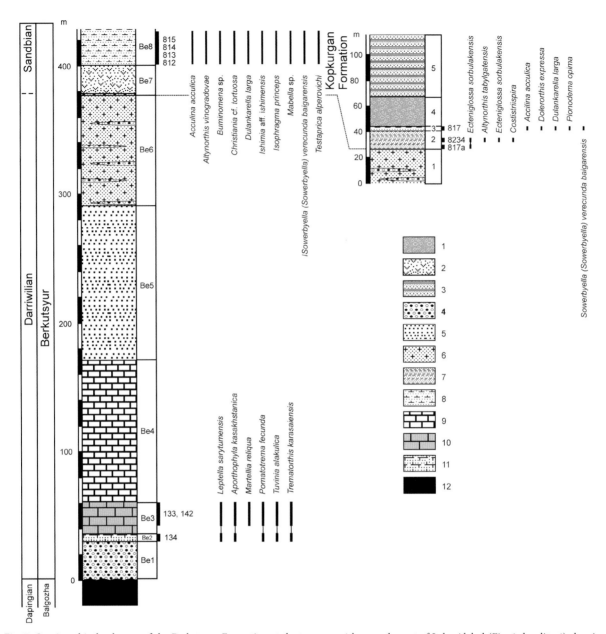

Fig. 7. Stratigraphical columns of the Berkutsyur Formation at the type area 4 km south-west of Lake Alakol (Fig. 1, locality 4) showing fossil localities and stratigraphical ranges of selected brachiopod species. Lithologies: 1. olistostrome horizons; 2, tuffaceous sandstones; 3, graded sandstones and siltstones; 4, polymict conglomerates; 5, coarse-grained, laminated sandstones; 6, laminated oligomict and arkosic sandstones; 7, intercalating sandstones and siltstones; 8, calcareous sandstones and siltstones; 9, thick-bedded and massive limestones; 10, nodular limestones with stromatolites and stromatoporoids; 11, polymict calcareous sandstones; 12, black siliceous shales.

24 m. Brachiopods are common throughout (localities 812–815).

The upper contact of the Berkutsyur Formation is seen south of the fault cutting the southern wing of the syncline (Figs 6, 7, localities 817, 817a, 8234). There the upper part is in ascending order: (1) medium to fine grained arkosic sandstones with lenticular beds of pebbly gritstones, more than 26 m; (2) greyish green to dark green, fine-grained sandstones intercalating with subsidiary siltstones, with lingulide shell beds in the lower part (Locality 817a) and rhynchonellide shell beds in the mid part (Locality 8234), 15 m; (3) greyish green to dark green, fine-grained sandstones and siltstones with lenticular brachiopod shell beds, 2 m; (4) dark grey to black argillites and siltstones with olistoliths of limestone, 24 m; (5) graded dark grey sandstones, siltstones and argillites over 50 m. The lowermost three units in the transect represent the upper part of the Berkutsyur

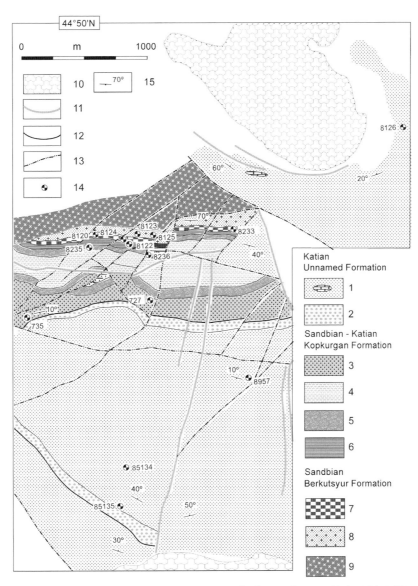

Fig. 8. Geological map of the area south of the Akzhartas Granitic Massif, 6 km south-west of Lake Alakol (Fig. 1, locality 5), showing position of fossil localities (modified from Popov *et al.* 2009). Lithologies: 1, 2, Late Katian (unspecified): 1, cross- and horizontal-laminated sandstones with occasional shell beds and carbonate build-ups; 2, polymict conglomerates; 3–6, Kopkurgan Formation (late Sandbian to early Katian): 3, quartzose sandstones with subsidiary conglomerates and brachiopod shell beds; 4, intercalating sandstones and siltstones; 5, olistostrome horizons; 6, laminated black graptolitic argillites; 7, 8, Berkutsyur Formation (Sandbian): 7, nodular limestones with dasyclad algae and bedded bioclastic limestones; 8, arkosic sandstones with lenses of calcareous siltstone and subsidiary conglomerates; 9, Middle Ordovician intrusive rocks of the Akzhal Complex; 10, salt marshes; 11, dykes; 12, angular unconformities; 13, faults; 14, fossil localities; 15, dip and strike.

Formation, while the lower boundary of the overlying Kopkurgan Formation is placed at the base of Unit 4, the lower olistolith horizon. The lower unit of arkosic sandstones is probably the equivalent of Unit Be6 of the previous section, but the overlying unit of tuffaceous sandstones is absent.

5.5–7 km further west, the Berkutsyur Formation is exposed on the south-west side of an unnamed salt lake where it transgressively overlies plagiogranites of the Akzhartas Granitic Massif assigned by Vinogradova (2008) to the Akzhal Complex. Here it comprises (Figs 8, 9): (1) brownish-yellow and greyish-brown arkosic sandstones with lenses of calcareous siltstone and with subsidiary polymict pebbly conglomerates and gritstones up to 25 m thick; (2) light-grey to dark grey nodular limestones with dasyclad algae and bedded bioclastic limestones, up to 10 m thick; (3) greenish-grey siltstones intercalating with nodular limestones, up to 4 m thick. The Kopkurgan Formation in the described transect is

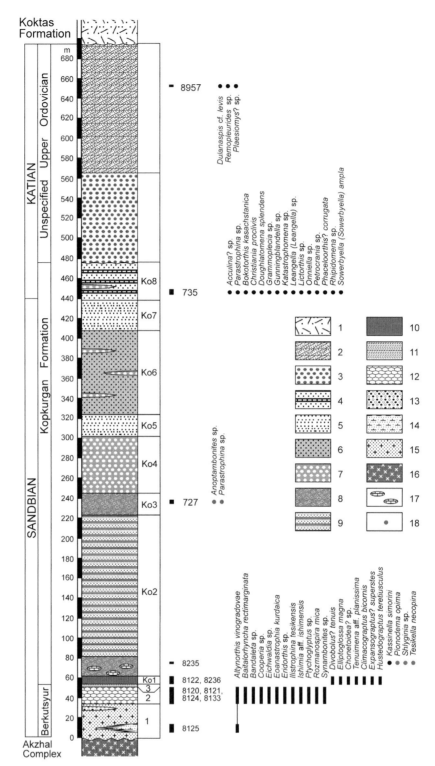

Fig. 9. Stratigraphical column of the Berkutsyur and Kopkurgan formations in the area south of the Akzhartas Granitic Massif, 6 km south-west of Lake Alakol (Fig. 1, locality 5), showing fossil localities and stratigraphical ranges of selected brachiopod species. Lithologies: 1, Koktas Formation (Devonian), volcanic rocks and tuffs; 2–3, Upper Ordovician, late Katian (unspecified): 2, cross- and horizontal-laminated sandstones with occasional shell beds; 3, polymict conglomerates; 4–10, Kopkurgan Formation (late Sandbian to early Katian): 4–8, intercalating sandstones and siltstones with subsidiary conglomerates and brachiopod shell beds; 5, quartzose sandstones; 6, sandstones with lenses of conglomerates and gritstones; 7, pebbly polymict conglomerates; 8, olistostrome horizons; 9, graded sandstones and siltstones; 10, black graptolitic argillites; 11–14, Berkutsyur Formation; 11, intercalating siltstones and nodular limestones; 12, nodular limestones with dasyclad algae; 13, pebbly polymict conglomerates; 14, calcareous siltstones; 15, arkosic sandstones; 16, volcanic rocks of the Akzhal Complex (Middle Ordovician); 17, large limestone olistoliths; 18, allochthonous fossil occurrences.

overlain by graptolitic black shales of the Berkutsyur Formation. The lowermost unit both in lithology and because of the occurrence of *Altynorthis vinogradovae* (Figs 8, 9; Locality 8125) is the equivalent of Unit 8 in the type section. Overlying nodular limestones contain the *Eoanastrophia* Association (Figs 8, 9: Localities 8120, 8120/4, 8121, 8124, 8233, 8236a).

Distribution. – The Berkutsyur Formation is exposed more or less continuously for about 50 km in the Sarytuma Fault Zone south-west of Lake Alakol and the western coast of Lake Balkhash. In the most complete section, 4 km south-west of Lake Alakol, it rests with a slight angular unconformity on the volcano-sedimentary Lower to Middle Ordovician Balgozha Formation. In the area 15 km west of Chimpek Bay and 6 km north-west of Sarytuma (Figs 1, 3), the Berkutsyur Formation is a succession of sandstones and siltstones with a thickness of over 100 m topped by a unit of grey algal and bioclastic limestones up to 50 m thick which is nodular in the lower part and thin to medium bedded in the upper part. Nodular limestones contain abundant calcareous algae including *Aphroporella*, *Coelosphaeridium*, *Solenopora*, and *Vermiporella* (identified by M. B. Gnilovskaya). A brachiopod shell bed in the uppermost part of the unit (Locality 388) contains abundant *Chaganella chaganensis* and a few *Glyptorthis* sp. Overlying argillites assigned to the Kopkurgan Formation (Locality 388a) contain *Meristopacha* sp.

Age. – The Darriwilian age of the Berkutsyur Formation is supported by indicative brachiopod genera such as *Aporthophyla*, *Leptella*, and *Martellia* characteristic of the *Martellia-Pomatotrema* Association (Nikitina *et al.* 2006). The Sandbian age of the upper part of the Berkutsyur Formation is mainly based on the brachiopods *Dulankarella larga*, *Esilia* cf. *E. tchetverikovae*, *Isophragma princeps*, and *Rozmanospira mica*, which are unknown in deposits older than the Sandbian Stage elsewhere in Kazakhstan (Popov & Cocks 2017).

Remarks. – The rocks here assigned to the Berkutsyur Formation were deposited in a forearc basin on the margin of an active volcanic arc (Popov *et al.* 2009). The carbonate and siliciclastic rocks including stromatolitic, algal and argillaceous limestones, occasional brachiopod shell beds, polymict conglomerates, arkosic and polymict horizontally and cross-laminated sandstones and siltstones showing retrogradation from the upper shoreface zone to the upper offshore. Characteristic brachiopod dominated assemblages are the lingulide (*Ectenoglossa* Association), early rhynchonellide (*Ancistrorhyncha* and *Baitalorhyncha* associations), and strophomenid dominated (*Acculina-Sowerbyella* and *Testaprica* associations) biofacies (BA1-3).

Kopkurgan Formation (New)

Derivation of name. – The name is derived from the Kopkurgan valley in the south-western part of the type area which is about 8.5 km south-west of Lake Alakol.

Stratotype. – The coordinates of the measured section base are 44°48′55″N, 74°01′58″E. and for the top of the section are 44°48′31″N, 74°2′6″E, and it is located on the northern flank of a syncline (Fig. 8), where it conformably overlies the Berkutsyur Formation.

Definition of boundaries. – In the sections south-west of Lake Alakol the Kopkurgan Formation rests conformably on the Darriwilian to early Sandbian Berkutsyur Formation and is overlain unconformably by an unnamed sedimentary unit of late Katian age. The base of the Kopkurgan Formation coincides with the maximum flooding surface, when deposition of black graptolitic shales (Unit Ko1) and olistostromes commenced throughout the region, suggesting tectonically-induced rapid subsidence caused by the formation of a foreland basin (Popov *et al.* 2009). It is followed by a coarsening-upward succession of graded sandstones and siltstones with conglomerate lenses and a second olistostrome horizon (Units Ko2 – Ko7). The succession ends with intercalating sandstone and siltstone units with horizontal lamination and hummocky cross-lamination and occasional brachiopod shell beds, suggesting a depositional environment transitional from the lower shore-face to upper offshore (Unit Ko8).

The Kopkurgan Formation is overlain by a minor angular unconformity with polymict pebbly conglomerates 40 m thick, succeeding a unit of parallel- and cross-laminated, coarse to medium grained sandstones with subsidiary conglomerates, gritstones and brachiopod shell beds, with an estimated thickness over 110 m, which form broad and shallow synclines in the southern and north-eastern part of the area south-west of Lake Alakol (Fig. 8). The Katian age of these deposits is proved by the brachiopod *Anaptambonites* sp. (Fig. 8; Locality 8126) and the trilobites *Dulanaspis* cf. *D. levis* and *Remopleurides* sp. (Figs 8, 9, Locality 8957). On the correlation charts of Nikitin (1991) and Popov *et al.* 2009, fig. 4) this unnamed unit was assigned to the Kyzylsai Formation, but that is not supported here because the Kyzylsai Formation, which is widely developed in the southern Chu-Ili Range, is an up to 1000 m thick succession of graded sandstones and siltstones with subsidiary conglomerates deposited in an extensional tectonic regime, while the unnamed late Katian unit developed south-west of Lake Alakol includes siliciclastic sedimentary rocks deposited mainly within the shoreface zone (Popov *et al.* 2009).

Description. – In the type section the succession in ascending order is as follows (Fig. 9):

Unit Ko1. – Laminated black graptolitic argillites, 8 m, with *Aploobolus tenuis* n. sp., *Chonetoidea?* sp., *Tenuimena* aff. *T. planissima*, *Elliptoglossa magna*, and trilobites (Figs 8, 9, Localities 8122 and 8236).
Unit Ko2. – Graded, greenish-grey sandstones and siltstones, with an olistostrome horizon at the base, up to 135 m. Matrix in the olistostrome (Fig. 8, Locality 8235) contains *Kassinella simorini*, *Pionodema opima*, *Tesikella necopina*, and *Shlyginia* sp.
Unit Ko3. – Matrix-supported pebbly, polimict conglomerate, with sandstones, and boulders of bioclastic limestone and calcarenite (second olistostrome horizon), total 18 m. Limestone olistoliths contain *Anoptambonites* sp. and *Parastrophina* sp. (Fig 8; Locality 727)
Unit Ko4. – Pebbly polymict conglomerate up to 30 m.
Unit Ko5. – Quartzose, medium-grained sandstones, 8 m.
Unit Ko6. – Coarse to medium-grained sandstones with lenses of conglomerates and gritstones, 35 m.
Unit Ko7. – Fine-grained quartzose sandstones up to 16 m.
Unit Ko8. – Sandstone, quartzose, medium to fine grained with brachiopod shell beds 0.5–1 m, 18–20 m. Locality 735 (Figs 8, 9), about 900 m west of the measured profile, contains a diverse brachiopod assemblage of the *Sowerbyella-Doughlatomena* Association.

Distribution. – The Kopkurgan Formation is exposed for about 50 km in the Sarytuma Fault Zone southwest of Lake Alakol and the western coast of Lake Balkhash.

Age. – The black shales at the base of the Kopkurgan Formation (Figs 8, 9; Locality 8122) contain graptolites identified by Damir Tsai as *Climacograptus bicornis*, *Expansograptus?* *superstes*, and *Hustedograptus teretiusculus*. The first two species indicate the lower part of the *Climacograptus bicornis* Biozone and a mid-Sandbian age. The moderately rich brachiopod fauna from the uppermost part of the Kopkurgan Formation includes *Bokotorthis kasachstanica*, *Christiania proclivis*, and *Sowerbyella (S.) ampla*, which also occur in the early to middle Katian Dulankara Formation of the southern Chu-Ili Range (Popov *et al.* 2000; Popov & Cocks 2006).

Area 15 km west of Chimpek Bay. – In the area 15 km west of Chimpek Bay and 6 km north-west of Sarytuma (Figs 1, 3), the Berkutsyur Formation is a succession of sandstones and siltstones with a thickness of over 100 m topped by a unit of grey algal and bioclastic limestones up to 50 m thick which is nodular in the lower part and thin to medium bedded in the upper part. Nodular limestones contain abundant calcareous algae including *Aphroporella*, *Coelosphaeridium*, *Solenopora*, and *Vermiporella* (identified by M. B. Gnilovskaya). A brachiopod shell bed in the uppermost part of the unit (Locality 388) contains abundant *Chaganella chaganensis* and a few *Glyptorthis* sp. Overlying argillites provisionally assigned to the Kopkurgan Formation (Locality 388a) contain *Meristopacha* sp.

Brachiopod associations

The transitional Mid to Late Ordovician deposits of the West Balkhash Region exhibit a wide spectrum of sedimentary environments varying from upper shore face to lower offshore on the shelves of the Chu-Ili Terrane, which represents a remnant of an Ordovician active ensialic island arc. The island shelves were inhabited by many benthic, often brachiopod-dominated associations, which show significant variations in taxonomic composition, richness, and abundance. Inferred depth relationships between different brachiopod associations are supported by the distinct retrogradational trends in the lower part of the Baigara Formation and within the transitional interval from the upper Berkutsyur to the lower Kopkurgan Formation. In the Chu-Ili Terrane there was in Late Darriwilian to early Sandbian times increased faunal turnover, local extinction, and immigration, marking the height of the global Great Ordovician Biodiversification Event, and making a significant impact on the Chu-Ili brachiopod communities.

The material used in the present study was collected over a period of 25 years and is variable in quality and not always well suited for quantitative analysis. It is based on 37 localities from the Baigara, Berkutsyur, Kopkurgan, Takyrsu, and

Age	S.S.	BA	Normal current activity		Organic build-ups, medium to high diversity	Generally quiet, affected by seasonal storms	Quiet water, disaerobic conditions may develop
			Low to medium diversity	Medium to high diversity			
Sandbian	Sa2	BA5					*Foliomena* Association
		BA4					
		BA3			*Parastrophina–Kellerella* Association		
					Acculina – Dulankarella Association		
		BA2	*Mabella – Sowerbyella* Association			*Adensu* Association	
			Tesikella Association				
		BA1	*Ectenoglossa* Association				
	Sa1	BA5					*Tenuimena - Elliptoglossa* Association
		BA4					
		BA3	*Chaganella* Association			*Eoanastrophia* Association	
			Acculina – Sowerbyella Association				
		BA2	*Testaprica* Association				
			Baitalorhyncha Association				
		BA1	*Ectenoglossa* Association				
Darriwilian	Dw3-Sa1	BA4					*Bimuria - Grammoplecia* Association
		BA3				*Altynorthis – Lepidomena* Association	
		BA2	*Ancistrorchincha* Association			*Sowerbyella – Altynorthis* Association	
		BA1					
	Dw1-2	BA5					Assemblage with *Akadyria*
		BA4	Assemblage with *Leptellina* and *Asperdelia*	*Taphrodonta* Association			*Metacamerella* Association
		BA3	Assemblage with *Leptellina*	*Martellia – Pomatotrema* Association			
		BA2	*Aporthophyla* Association				
			Pseudolingula – Kopella Association				
		BA1					

Fig. 10. Darriwilian to Sandbian community framework for the Chu-Ili, Betpak-Dala, and West Balkhash regions.

Tastau formations, combined with analogous data from the late Sandbian Anderken Formation published by Popov *et al.* (2002) and from the Darriwilian Uzunbulak Formation by Nikitina *et al.* (2006). Analysis of taxonomic composition and relative abundance of brachiopod taxa from many localities in the Anderken Formation allows recognition of the associations characterised below which are interpreted within the Benthic Assemblage (BA) scheme of Boucot (1975).

We can recognise ten associations from the Berkutsyur and Baigara formations: *Acculina-Sowerbyella* Association, *Altynorthis-Lepidomena* Association, *Ancistrorhyncha* Association, *Baitalorhyncha*

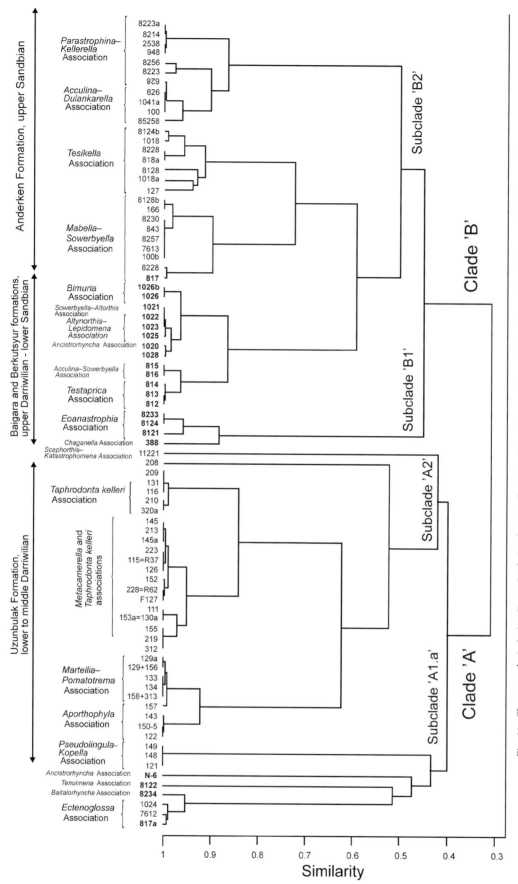

Fig. 11. Cluster analysis (using Raup-Crick Similarity) of 120 brachiopod genera from 88 localities of Darriwilian to Sandbian age from the Chu-Ili Range, West Balkhash Region, and Betpak-Dala. Data for the Uzunbulak Formation are from Nikitina *et al.* (2006). Data for the Anderken Formation are from Popov *et al.* (2002). Localities in the Baigara, Berkutsyur, and Kopkurgan formations are highlighted in bold.

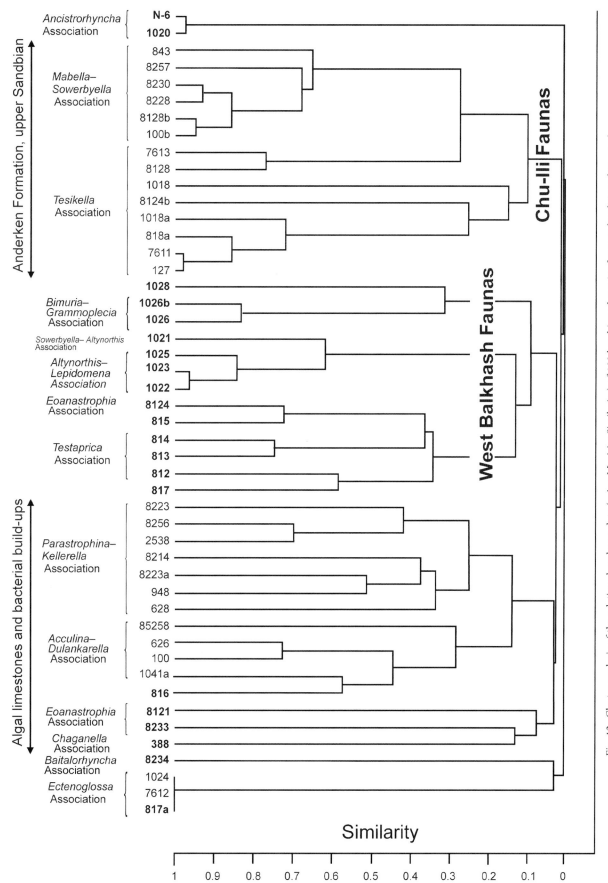

Fig. 12. Cluster analysis of the relative abundance data (using Morisita Similarity) of 102 brachiopod species from 48 localities of Darriwilian to Sandbian age from the Chu-Ili Range, West Balkhash Region, and Betpak-Dala. Data for the Anderken Formation are from Popov *et al.* (2002). Localities from the Baigara, Berkutsyur, and Kopkurgan formations are highlighted in bold.

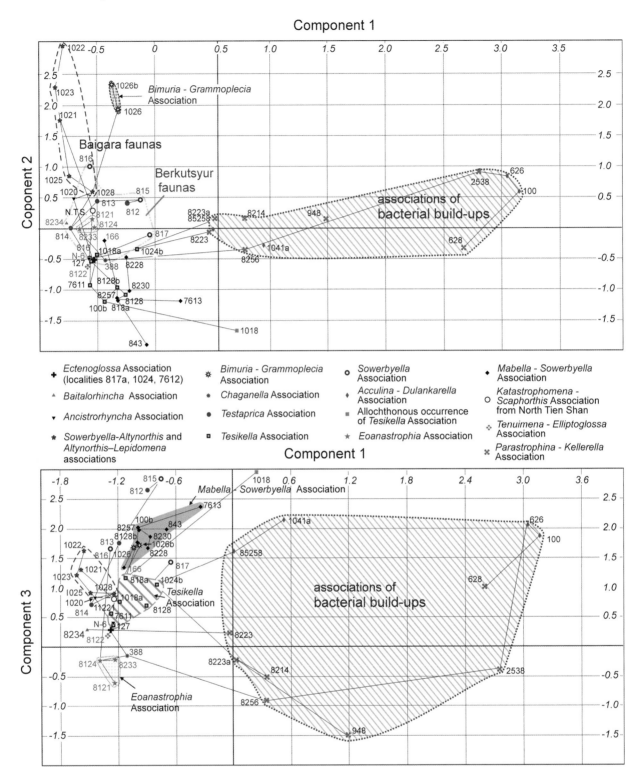

Fig. 13. Two-dimensional Principal Components Analysis plots on first, second and third eigenvectors of 87 brachiopod genera from 49 localities of late Darriwilian to Sandbian ages from the Chu-Ili Range, West Balkhash Region, and Betpak-Dala. Data for the Anderken Formation are from Popov *et al.* (2002).

Association, *Bimuria-Grammoplecia* Association, *Chaganella* Association, *Ectenoglossa* Association, *Eoanastrophia* Association, *Sowerbyella-Altynorthis*

Association, and the *Testaprica* Association; and three more associations from the Kopkurgan Formation: *Tenuimena-Elliptoglossa* Association, *Tesikella*

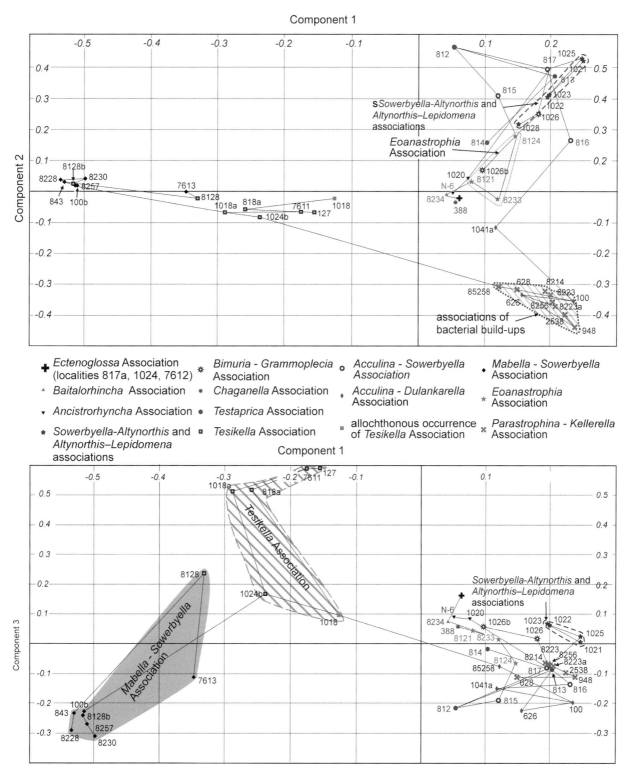

Fig. 14. Two-dimensional Principal Coordinates Analysis plots on first, second, and third eigenvectors of 102 brachiopod species from 48 localities of late Darriwilian to Sandbian age from the Chu-Ili Range, West Balkhash Region, and Betpak-Dala. Data for the Anderken Formation are from Popov *et al.* (2002).

Association, and the *Sowerbyella- Doughlatomena* Association (Fig. 10). Primary numerical data on the brachiopod occurrences and abundance (listed in the outline of the localities) were analysed using

the Simpson dominance (λ), Buzas and Gibson's evenness (e^H/S) and Margalef's richness (d) indices. Simpson Dominance (λ) is calculated as $k = \Sigma_i(n_i/n)^2$, where n is total number of individuals and n_i

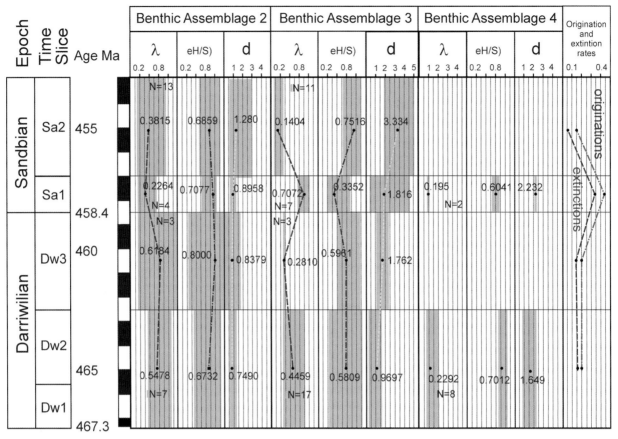

Fig. 15. Variations in generic diversity (dominance, evenness, and Margalef's richness indices) for the Darriwilian to Sandbian brachiopod communities (BA2–4) from the Betpak-Dala, West Balkhash and Chu-Ili regions, and comparative origination/extinction rates (lineage per million years; Lma) for the Darriwilian to Sandbian of the Chu-Ili Terrane. Variation in scores of individual brachiopod associations within a particular time slice are shown in grey, while average scores are centred against individual time slices. Comparative origination and extinction rates for Sandbian to Katian rhynchonelliform genera from selected Kazakh terranes. The Lma for origination and extinction rates (y axis) is plotted as the number of generic originations (or extinctions) within the particular chronostratigraphical time interval, divided by the total generic diversity within the unit, divided by the chronological duration of the interval (Patzkowsky & Holland, 1997); calculations for Lma are plotted at the midpoint of each time interval.

is total number of individuals of taxon i. Buzas and Gibson's evenness is calculated as e^H/S, where H (Shannon index) is calculated as $H = -\Sigma_i \, n_i/n$. ln n_i/n. Margalef's richness index (d) is calculated as $d = (S - 1)/\log_e n$, where S is number of species.

The distribution of 120 genera from 26 localities from the Baigara, Berkutsyur, Kopkurgan, and Tastau formations plus 62 localities from the Anderken, Rgaity, and Uzunbulak formations was analysed by Cluster Analysis (Raup-Crick similarity indexes). The same matrix (excluding samples from the Uzunbulak Formation and Alakol Limestone) was subjected to Principal Components Analysis (Figs 11, 13). The second relative abundance matrix for 102 species, in 21 localities from the Baigara and Berkutsyur Formations in the West Balkhash Region plus about 1,800 brachiopod specimens from 27 localities from the Anderken Formation of the South Chu-Ili Range, was subjected to Cluster Analysis

(Morisita similarity indexes) and Principal Coordinate Analysis depicting the Morisita similarities (Fig. 14). The Morisita Index was chosen because it was developed for the analysis of relative abundance data and is reasonably independent of sample size effects. All calculations have been made using the PAST statistical package (Hammer & Harper 2006).

The Ectenoglossa *Association*

This is a monospecific linguloid BA1 association ($\lambda = 1$, $e^{H/S} = 1$, d = 0, N = 4, including localities from the Anderken Formation) characteristic of coarse to fine grained mobile sands with subhorizontal and cross stratification deposited within the shoreface zone (Fig. 10). It is common in the late Sandbian Anderken Formation (Popov *et al.* 2002) and from the single Locality 817a in the upper part of the Berkutsyur Formation, where lingulides occur as accumulations of complete disarticulated valves on the bedding surfaces of subhorizontally laminated fine grained sandstones where no complete shells are preserved *in situ*; but in the Anderken Formation *Ectenoglossa sorbulakensis* shells fossilised within their burrows are sometimes present, and gastropods and bivalves also occur there. Outside the Chu-Ili Terrane the *Ectenoglossa* Association is also known from the Bestamak Formation (Sandbian) of the Chingiz–Tarbagatai island arc system (Popov & Cocks 2017), and in the North Tien Shan Microplate the Sandbian *Tunisiglossa almalensis* Association probably inhabited a similar environment (Popov *et al.* 2007). In the cluster analyses of the Sandbian faunas of the Chu-Ili Terrane, the *Ectenoglossa* Association forms a distinct subcluster (Fig. 12; Subclade 'A1.a') together with the Darriwilian to Sandbian associations of rhynchonellide biofacies and the early to mid Darriwilian *Pseudolingula-Kopella* Association. The Morisita Similarity to other faunas approaches zero (Fig. 12). In the Principal Components Analysis, the samples from the *Ectenoglossa* Association had negative values along the first two maximum direction variations and low positive scores along Axis 3 (Fig. 13), while in the Principal Coordinate Analysis (Morisita Similarity) it shows low positive scores in the first and third major directions of variation and low negative scores along the second direction of variation (Fig. 14), which place them close to the contemporaneous rhynchonellide associations.

The Ancistrorhyncha Association

This late Darriwilian association (λ = 0.702, $e^{H/S}$ = 0.462–1, d = 0; N = 2) is the earliest example of the rhynchonellide biofacies of BA2 (Fig. 10) yet documented from anywhere in the world. The rhynchonellide *Ancistrorhyncha modesta* together with bivalve molluscs form dense accumulations of disarticulated valves in shell beds within units of coarse to fine grained sandstones in the lower Baigara Formation (Locality 1020) and the middle Berkutsyur Formation (Locality N-12). While the *Ancistrorhyncha* shells are allochthonous, they can only have lived in a monospecific brachiopod association in a high energy nearshore environment affected by seasonal storms. *Altynorthis betpakdalensis, Katastrophomena rukavishnikovae*, and *Sowerbyella (S.) verecunda baigarensis* also occur sporadically in the rhynchonellide shell beds of the Baigara Formation, but apparently not in association with *Ancistrothyncha modesta* and were probably derived from adjacent brachiopod communities, probably the *Scaphorthis-Katastrophomena* Association. When a Morisita Similarity index, focused on taxonomic composition and

relative abundance of brachiopod species, is applied, the *Ancistrorhyncha* Association is very dissimilar to all the other late Darriwilian to Sandbian faunas (Fig. 12). In generic composition (presence and absence data), analysed applying Raup-Crick Similarity, a 'pure' sample of the *Ancistrorhyncha* Association from the lower Berkutsur Formation (Locality N-6) groups together with other samples from lingulide and rhynchonellide biofacies characteristic of BA1-BA2 (Fig. 11; Subclade 'A1.a'). Another sample (from Locality 1020) groups closely with the strophomenide-plectorthid association, probably because of the sporadic presence of *Altynorthis* and *Sowerbyella (S.)* in the rhynchonellide shell beds from the Baigara Formation and by taphonomic bias. In the Principal Components Analysis, the samples from the *Ancistrorhyncha* Association show moderately negative scores along Axis 1, and variable, low positive to moderately negative scores along Axis 2 and low positive scores along the third maximum direction variations (Fig. 13). In the Principal Coordinate Analysis (Morisita Similarity) the samples of the *Ancistrorhyncha* Association show low positive scores along the first and third maximum direction variations, while the scores along the second direction of variation approach zero (Fig. 14).

The Baitalorhyncha Association

This BA2 association (λ = 0.447, $e^{H/S}$ = 0.507, d = 1.00, N = 1) is known only from Locality 8234 in the upper Berkutsyur Formation, in a shell bed within a unit of intercalating fine grained sandstones and siltstones. The shells are accumulations of complete disarticulated valves, which occur with small unidentified gastropods. The rhynchonellide *Baitalorhyncha rectimarginata* (48%), and the early lissatrypoid *Costistriispira proavia* (46%) are the two dominant taxa and are accompanied by *Altynorthis tabylgatensis, Ectenoglossa sorbulakensis*, and *Trematis*? sp. While the shell bed at Locality 8234 was probably formed during a seasonal storm event, the presence of relatively complete thin shelled valves of linguliform brachiopods, which cannot withstand long-distance transportation (Emig 1997), and the absence of abrasion on the rhynchonellide and atrypide shells, suggest that they were preserved near their original habitat, which was probably a rimmed shallow shelf (Fig. 10). This is the earliest mixed rhynchonellide – atrypide association yet documented from anywhere. The closest analogue is the *Altaethyrella* [=*Lydirhyncha*]-*Nalivkinia* (*Pronalivkinia*) [=*Rongatrypa*] Association described by Popov *et al.* (2000) from the subsequent Otar Beds (early Katian) in the southern Chu-Ili Terrane.

In generic composition, when presence/absence data are analysed using Raup-Crick Similarity, the *Baitalorhyncha* Association is dissimilar to all other late Darriwilian to Sandbian brachiopod associations, with a score approaching zero (Fig. 11), but tends to group with the lingulide *Ectenoglossa* Association. At the species level the *Baitalorhyncha* Association also grouped with the *Ectenoglossa* Association) showing medium similarity (Fig. 10), probably because of the occasional occurrence of *Ectenoglossa sorbulakensis*, while other core taxa (*Baitalorhyncha* and *Costistriispira* are exclusively confined to this association. In the Principal Components Analysis the *Baitalorhyncha* Association shows high negative scores along the first and low positive scores along the second and third maximum direction variations.

The Tesikella *Association*

The brachiopods within this low richness BA2 association were described by Popov *et al.* (2002) from the middle Anderken Formation (upper Sandbian). The association is widespread within the Chu-Ili terrane, but although it does not occur *in situ* in West Balkhash, the occasional presence of *Tesikella necopina* in the olistostrome horizon at the base of the Kopkurgan Formation (Locality 8235) must be an indication of a *Tesikella* Association nearby. *Tesikella necopina* has a narrow stratigraphical range confined to the late Sandbian, and is endemic to the Chu-Ili Terrane.

The Sowerbyella-Altynorthis *Association*

There are various medium rich and high density associations, probably of late Darriwilian age and consisting of strophomenides and plectorthids, which inhabited quiet nearshore environments on a rimmed carbonate shelf. Among them, the *Sowerbyella-Altynorthis* Association (λ = 0.1535, $e^{H/S}$ = 0.6875, d = 1.758, N = 1) is known from a single locality in the lower part of the Baigara Formation. There are no distinct dominants, but the most common taxa are *Altynorthis betpakdalensis* (20.7%), *Leptellina* sp. (20.3%), *Sowerbyella* (S.) *verecunda baigarensis* (19.3%), and *Acculina acculica* (10.8%), which together make up more than 70% of the total individuals. Other common species include *Katastrophomena rukavishnikovae* (8.8%), *Colaptomena insolita* (7.8%), and *Scaphorthis recurva* (7.5%). Minor components are *Bimuria* sp., *Ectenoglossa sorbulakensis*, *Plectocamara* sp., and *Sonculina baigarensis*, which are together less than 6% of the total individuals. Only *Plectocamara* is exclusive to the *Sowerbyella-Altynorthis* Association; but the abundance of *Acculina*,

Katastrophomena, Leptellina, Sowerbyella (S.), and *Scaphorthis* significantly declined in the *Lepidomena-Altynorthis* Association, which replaced the *Sowerbyella-Altynorthis* Association in a transgressive succession. The proliferation of plectambonitoids including *Sowerbyella* (S.) *verecunda baigarensis* is also characteristic of the early Sandbian *Sowerbyella* Association, and that subspecies together with *Acculina acculica* are also distinctive components of the *Testaprica* Association. Nevertheless, these two associations are somewhat distant from the *Sowerbyella-Altynorthis* Association in the Raup-Crick and Morisita cluster analyses. While *Acculina, Katastrophomena*, and *Scaphorthis* are common and represented by the same species in the approximately contemporaneous *Scaphorthis-Katastrophomena* [= *Strophomena*] Association from the North Tien Shan Microcontinent (Nikitina 1985; Popov & Cocks 2017) which inhabited a shallow clastic shelf (BA2), the latter differs in low richness and in the occurrence of *Paralenorthis* and *Oepikina*, neither of which are present in the late Darriwilian associations of Chu-Ili. In the Raup-Crick Cluster Analysis the *Scaphorthis-Katastrophomena* Association shows no distinct similarity with all the other Chu-Ili faunas (Fig. 11), which is in good agreement with the analysis of the Kazakh late Darriwilian to early Sandbian faunas by Popov & Cocks (2017, fig. 7).

The Altynorthis-Lepidomena *Association*

This association (λ = 0.2643–0.2728, $e^{H/S}$ = 0.4042–0.6795, d = 1.758–0.2413, N = 2) is known from three localities (1022, 1023 and 1025) in the lower Baigara Formation, and occurs in slightly argillaceous limestones rich in dasyclad algae in association with unidentified bryozoans, a few trilobites, and orthocone cephalopods. The brachiopod shells are largely articulated and occasionally in life position, suggesting that the samples came from the area originally inhabited by this association. *Altynorthis betpakdalensis* is the most common species (43–52% of total individuals) and is ubiquitous. Another indicative species is *Lepidomena betpakdalensis* (10–15%), which is confined to the *Altynorthis-Lepidomena* Association, together with such taxa as *Apatomorpha akbakaiensis, Eremotoechia inchoata, Grammoplecia* aff. *G. globosa, Multispinula* sp., and *Trematis* aff. *T. parva*, which together comprise 13–35% of individuals in the community. In the Morisita Cluster Analysis this association forms a small subcluster with similarity levels over 0.8 (Fig. 12), but in the Raup-Crick Cluster Analysis it cannot be distinguished from the *Sowerbyella-Altynorthis* Association (Fig. 11), probably because a large proportion of ubiquitous taxa, including *Acculina acculica*, *Colaptomena insolita*, *Leptellina* sp., *Katastrophomena rukavishnikovae*, *Scaphorthis recurva*, *Sonculina baigarensis*, and *Sowerbyella (S.) verecunda baigarensis*. In the Principal Components Analysis (genus level similarity), the *Sowerbyella-Altynorthis* and *Altynorthis-Lepidomena* associations cluster closely together without significant overlap with the other Sandbian faunas, showing high negative scores along Axis 1 and moderately high positive scores along the second and third maximum direction variations (Fig. 13). They are distinctly separate from the late Sandbian *Parastrophina-Kellerella* and *Acculina-Dulankarella* associations characteristic of the bacterial build-ups. The latter are the only associations with positive scores of the first maximum direction variations in the analysis.

In the Principal Coordinate Analysis (Morisita Similarity) the *Sowerbyella-Altynorthis* and *Altynorthis-Lepidomena* associations also group closely (Fig. 14), showing variable positive scores for all three-first maximum direction variations, while approaching zero levels along the third major direction variation. Although they overlap significantly with early Sandbian brachiopod associations in the first and second first maximum direction variation scores, they are clearly different from the late Sandbian brachiopod faunas of the shallow clastic shelf (*Tesikella* and *Mabella-Sowerbyella* associations), which show medium to high negative first maximum direction variation scores, while the late Sandbian brachiopod associations of the organic build-ups exhibit high negative scores along the second first maximum direction variation.

The Testaprica *Association*

This is a low to medium diversity association (λ = 0.1503–0.4587, $e^{\wedge H/S}$ = 0.7014–0.8461, d = 0.6470–0.2090, N = 3) defined by Popov & Cocks (2017) and documented from three localities (812, 813, and 814) in the upper Berkutsyur Formation, where the brachiopods are predominantly disarticulated valves without significant breakage or abrasion, dispersed in the sediment, or as loose lenticular shell accumulations, and were probably preserved near their original habitat. It was a quiet environment, occasionally affected by seasonal storms, in a shallow, restricted clastic nearshore shelf (BA2) (Fig. 10). The commonest species are *Testaprica alperovichi* (11–45%), which is entirely confined to this association, *Altynorthis vinogradovae* (6.5–50%), and *Ishimia* aff. *ishimensis* (4.6–21%), which are also minor components in the *Acculina-Sowerbyella* and *Eoanastrophia* associations. The taxonomic composition and

relative proportions of other brachiopod taxa, including *Buminomena* sp., *Christiania* cf. *C. tortuosa*, *Dulankarella larga*, *Isophragma princeps*, *Mabella* sp., and *Sowerbyella (S.) verecunda baigarensis*, is very variable between localities. In the Raup-Crick Cluster Analysis, samples of the *Testaprica* Association are placed closely together within the same subclade (Fig. 11; Berkutsyur Subclade) as the *Acculina-Sowerbyella* and *Ancistrorhyncha* associations. Similarity with the *Ancistrorhyncha* Association is possibly due to the co-occurrence of *Altynorthis* and *Sowerbyella (S.)*. In the Morisita Cluster Analysis, samples assigned to the *Testaprica* Association cannot be discriminated confidently from those of the *Acculina-Sowerbyella* Association (Fig. 11).

The Acculina-Sowerbyella *Association*

This is a low to medium rich BA3 association (λ = 0.1950–0.4757, $e^{H/S}$ = 0.5871–0.7922, d = 1.259–2.352, N = 3) [= *Acculina* Association of Popov & Cocks 2017] and is confined to the upper part of the Berkutsyur Formation (Localities 815, 816, and 817) where it occurs in variably calcareous siltstones and argillites (Fig. 10). Associated fossils include unidentified gastropods, trilobites (represented by *Pliomerina* sp. and unidentified asaphids), and occasional dasyclad algae. The brachiopods are usually in dense concentrations, but are mainly dispersed and unbroken disarticulated valves without abrasion. The proportion of articulated shells varies from 16% to 39%. The brachiopods were probably preserved within the general area of their original habitat, which was a shallow restricted shelf offshore within the euphotic zone in the retrogradational succession of the uppermost Berkutsyur Formation 4 km south-west of Lake Alakol (Figs 6, 7). The *Acculina-Sowerbyella* Association succeeds the *Ectenoglossa* (BA1) and rhynchonellide-lissatrypoid *Baitalorhyncha* associations (BA2). There are no distinct dominant taxa in the association, except *Sowerbyella (S.) verecunda baigarensis* (10–67%), and the proportion of other species varies between localities. As well as *Acculina acculica* the most common taxa are *Ishimia* aff. *I. ishimensis* and *Dulankarella larga*. While some brachiopods are shared with the *Testaprica* Association, including *Altynorthis vinogradovae*, *Buminomena* sp., *Isophragma princeps*, and *Mabella* sp. they represent minor components of the fauna, and *Testaprica* itself is completely missing. The brachiopod assemblage from Locality 816 is somewhat different in the presence of *Bandaleta* sp., *Esilia* cf. *E. tchetverikovae*, and *Sowerbyella (Sowerbyella)* cf. *S. acculica*, which are also characteristic of the *Eoanastrophia* Association. Nevertheless in the Raup-Crick Cluster Analysis it is placed in a single subcluster with other samples here assigned to the *Acculina-Sowerbyella* Association (Fig. 12), while in the Morisita Analysis it appears in the same subcluster as samples from the late Sandbian Anderken Formation assigned to the *Acculina-Dulankarella* Association (Fig. 11). In the Principal Components Analysis the areas occupied by the *Testaprica* and *Acculina-Sowerbyella* associations significantly overlap, showing negative scores along Axis 1, slightly negative to slightly positive scores along Axis 2, and moderately to highly positive scores along Axis 3 (Fig. 13). In the Principal Coordinate Analysis, the *Testaprica* and *Acculina-Sowerbyella* associations also cannot be discriminated, but in general their samples are characterised by higher positive first maximum direction variation and more negative second maximum direction variation scores by comparison with the contemporaneous *Eoanastrophia* Association.

The Eoanastrophia *Association*

This medium rich BA3 brachiopod association (λ = 0.2514–0.2866, $e^{H/S}$ = 0.5620–0.8190, d = 1.276–3.827, N = 3) is known from nodular limestones with abundant dasyclad algae in the uppermost part of the Berkutsyur Formation at three localities (8120, 8124, and 8233). It differs from other early Sandbian brachiopod associations from the Chu-Ili Terrane in the constant presence of camarelloid pentamerids, among which *Eoanastrophia kurdaica* is the most common, and occurs at all three localities. Other taxa confined to the association are *Cooperea* sp., *Eichwaldia* sp., *Ilistrophina tesikensis*, *Paraoligorhyncha*? sp., *Ptychoglyptus* sp., *Rozmanospira mica*, and *Synambonites* sp. Almost all of them are rare apart from the lissatrypoid *Rozmanospira*, which is relatively common at the single Locality 8121. Other taxa, which occur in the assemblage, include *Altynorthis vinogradovae*, *Baitalorhyncha rectimarginata*, *Bandaleta* sp., *Esilia* cf. *E. tchetverikovae*, *Ishimia* aff. *I. ishimensis*, and *Triplesia* sp. The high proportion of the articulated shells (55–65%) in the samples suggests that they are preserved within the area of their original habitat within a shallow shelf offshore, which was apparently a lime mud rich in bioclasts. In the Raup-Crick Cluster Analysis the *Eoanastrophia* Association shows little similarity (below 0.4) with other late Darriwilian to Sandbian faunas and forms a separate first order subcluster together with the *Chaganella* Association (Fig. 12). A small sample from the Kipchak Limestone at North Betpak-Dala (Locality 166) does not contain pentamerides; nevertheless, it appears to be in close

association with the *Eoanastrophia* Association in the Raup-Crick Cluster Analysis. In the Morisita Cluster Analysis the samples assigned to the *Eoanastrophia* Association form a separate subcluster together with the *Chaganella* Association within the first order subcluster uniting the West Balkhash faunas (Fig. 11), with the exception of the sample from Locality 8124 which shows closer similarity to those from the *Testaprica* and *Acculina-Sowerbyella* associations. In the Principal Components Analysis, samples assigned to the *Eoanastrophia* Association differ from all other late Darriwilian to Sandbian associations in having negative scores along the third maximum direction of variation (Fig. 13). In the Principal Coordinate Analysis the differences of the *Eoanastrophia* Association from other contemporary faunas are less evident, yet they show consistently lower positive scores along the second major axis of variation (Fig. 14). Outside the Chu-Illi Terrane the *Eoanastrophia* Association can be compared with the approximately contemporaneous *Rozmanospira* Association of the Chingiz-Tarbagatai island arc system. Both are characterised by the co-occurrence of *Rozmanospira*, *Eridorthis*, and *Esilia* and the presence of camarelloids, but they were placed some distance apart in the Raup-Crick Cluster Analysis of the Kazakh early Sandbian faunas by Popov & Cocks (2017).

The Chaganella *Association*

This is an oligotaxic brachiopod association recovered from a lenticular shell bed on top of a limestone with abundant calcareous green and red algae at the single Locality 388. It inhabited environments within the euphotic zone of restricted shallow carbonate shelf (BA 3). The plectambonitoid *Chaganella chaganensis* is the dominant taxon in the association, which also includes an unnamed species of *Eridorthis* as the only other component. *Chaganella chaganensis* was previously unknown in Chu-Ili, while in the Chingiz-Tarbagatai island arc system it occurs as a secondary component in the BA2–3 *Palaeotrimerella* Association (Popov & Cocks, 2017). In the Raup-Crick cluster analysis the *Chaganella* Association appears in a single second order subcluster together with the *Eoanastrophia* Association with similarity levels exceeding 0.8 (Fig. 12), probably due to the co-occurrence of *Eridorthis*; while in the Morisita Cluster Analysis its similarity with samples of the *Eoanastrophia* Association drops below 0.2 (Fig. 11). In the Principal Components Analysis the *Chaganella* Association shows slightly negative scores in all three maximum direction variations (Fig. 13) and

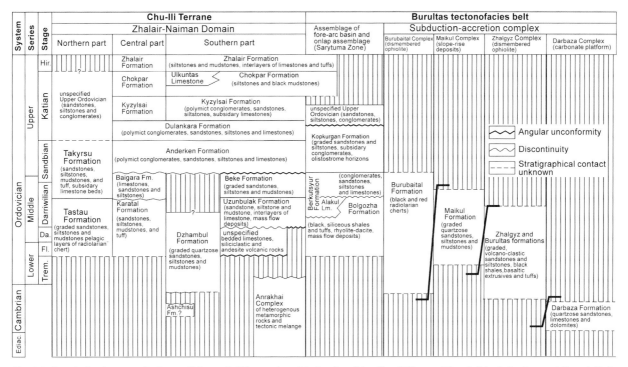

Fig. 16. Correlation between the Lower Palaeozoic lithostratigraphical units of the Chu-Ili Range, West Balkhash Region, and Betpak-Dala (much modified from Popov *et al.* 2002; 2009).

occupies a position on the margins of the area of the *Eoanastrophia* Association.

The Bimuria-Grammoplecia *Association*

This BA4 brachiopod association (λ = 0.1775–0.2175, $e^{H/S}$ = 0.5235–0.6847, d = 2.118–2.346, N = 2) is known from two localities (1026, 1026b) in the middle part of the Baigara Formation. Its original habitat was probably in patches of carbonate mud rich in bioclasts, which developed in the area of accumulation of fine siliciclastic, mainly silty sediments offshore, below the seasonal storm wave base (Fig. 10). Remarkably, articulated brachiopod shells (up to 95%) are confined to the lenticular beds of silty carbonates (Locality 1026b), while they occur largely disarticulated in the background siliciclastic deposits (Locality 1026). *Grammoplecia globosa* (38.1–37.5%) and *Bimuria* (9.7–22%) are the most common species in the association. Other characteristic taxa are *Christiania* cf. *C. tortuosa*, *Altynorthis tabylgatensis*, *Eremotoechia spissa*, and *Sowerbyella* (*S.*) *verecunda baigarensis*, which together comprise 33–43% of the total number of individuals in the assemblage. Other taxa, including *Atelelasma* sp., *Kajnaria derupta*, *Pseudocrania karatalensis*, *Sonculina* cf. *S. prima*, *Titanambonites* cf. *T. magnus*, and *Trematis* aff. *T. parva*, are rare. In the Raup-Crick Cluster Analysis the *Bimuria-*

Grammoplecia Association appears within the Baigara Subclade (Fig. 12) with similarity score to other faunas from the Baigara Formation slightly exceeding 0.75. This is probably because of the co-occurrence of a significant number of genera, including *Altynorthis*, *Bimuria*, *Eremotoechia*, *Grammoplecia*, *Sonculina*, *Sowerbyella* (*Sowerbyella*), and *Trematis*. In the Morisita Cluster Analysis, when relative abundance and species composition is taken into account, the *Bimuria-Grammoplecia* Association shows very weak links (similarity scores slightly below 0.1) with other approximately contemporaneous West Balkhash faunas apart from the *Eoanastrophia* Association (Fig. 11). In the Principal Components Analysis, the samples from the *Bimuria-Grammoplecia* Association are placed close together with moderately negative scores along Axis 1, and high positive scores along Axis 2 outside the area occupied by other late Darriwilian to Sandbian brachiopod associations (Fig. 13). The *Bimuria-Grammoplecia* Association can be compared with the early Sandbian *Bimuria-Kajnaria* Association of the Chingiz-Tarbagatai island arc system (Popov & Cocks 2017, Text-fig. 7). *Bimuria* and *Kajnaria* are common in both associations. However, the *Bimuria-Grammoplecia* Association shows higher diversity and includes strophomenides, triplesiidines, orthides, and craniides, while the *Bimuria-Kajnaria* Association is exclusively plectambonitoids.

Deep water brachiopod associations

The black graptolitic argillites at the base of the Kop-kurgan Formation (early Sandbian) contain low diversity brachiopods assigned to the *Tenuimena-Elliptoglossa* Association, best represented at Locality 1022. It includes *Tenuimena* aff. *T. planissima* and *Elliptoglossa magna* as the most common taxa, while other associated brachiopods are the obolid *Aploobolus? tenuis* and the plectambonitoid *Chonetoidea?* sp. There is also the unusual occurrence of a single specimen provisionally assigned to *Costistriispira?* sp., the early spire-bearer which is otherwise confined to the nearshore rhynchonellide biofacies. The associated fossil assemblage (in addition to graptolites) includes a moderately rich trilobite fauna (e.g. *Robergia, Nanshanaspis, Nileus, Pseudoampyxina,* and *Telephina*), occasional gastropod shells, and machaeridian sclerites. The association inhabited basinal environments below the photic zone and maximum storm wave base (BA5). It shows no interaction with other Sandbian brachiopod associations and was therefore not included in the cluster analysis; however, at Chu-Ili *Chonetoidea* and *Tenuimena* are common in the Darriwilian medium diverse – low dominance *Metacamarella* Association (BA4–5) of Nikitina *et al.* (2006), but in comparison to the latter the generic diversity in the *Tenuimena-Elliptoglossa* Association decreased considerably, probably due to the disaerobic condition prevailing at the water-sediment interface.

The pelagic layers of radiolarian cherts and siliceous argillites of the Tastau Formation (Darriwilian) in the upper part of the individual Bouma cycles (localities 154, 162, and 163) occasionally contain the linguliform brachiopods *Broeggeria* cf. *B. putilla, Oxolosia?* sp., and, remarkably, the rhynchonelliform brachiopod *Akadyria simplex.* Unlike the linguliform brachiopods, which spread to the environments of the abyssal plain by the end of the Early Ordovician (Tolmacheva *et al.* 2004), probably, as a part of symbiotic brachiopod-hexactinellide sponge benthic associations (Mergl 2002; Holmer *et al.* 2005), the occurrence of *Akadyria simplex* is the earliest of any rhynchonelliform brachiopod in a bathyal environment and the only Ordovician record. An allochthonous explanation due to distant transport across the slope or an epiplanktonic habit are both unlikely, since these minute plectambonitoid shells lack pedicle attachment and are often preserved articulated in the pelagic sediment. *Akadyria simplex* was originally described by Nikitina *et al.* (2006) from the graptolite-bearing argillites of the Uzunbulak Formation (Darriwilian) in the southern Chu-Ili Range, where it occurs with the linguliform brachiopods *Spondyglossella* sp., *Broeggeria* cf. *B. putilla,* and *Talasotreta* aff. *T. gigantea,* probably within BA5 or BA6.

The Sowerbyella-Doughlatomena *Association*

This is the only Katian brachiopod association ($\lambda = 0.1417$, $e^{H/S} = 0.677$, $d = 2.606$, $N = 1$) known from the West Balkhash Region. It occurs in a shell bed at Locality 735 in the upper part of the Kopkurgan Formation, representing possibly a proximal tempestite formed during a strong storm event. While most of the specimens occur as disarticulated valves (94%), they show little breakage and no significant abrasion, and thus the shell bed must have been formed relatively close to its original habitat, which was probably within the clastic shelf of the upper offshore (BA3). While a proportion of the taxa in this moderately rich allochthonous assemblage was somewhat distorted due to transportation within the water column, there is no reason to conclude that the species composition was not close to that of the original life association. The *Sowerbyella-Doughlatomena* Association is most comparable to the approximately contemporaneous *Lictorthis* [=*Plectorthis*]-*Metambonites* Association from the Degeres Beds (mid Katian) of southern Chu-Ili in the presence of such genera as *Anoptambonites, Christiania, Gunningblandella, Lictorthis, Leangella, Lydirhyncha, Rhipidomena,* and *Sowerbyella (Sowerbyella).* However, it lacks the strophomenoid genera *Glyptomenoides, Platymena,* and *Strophomena,* which are replaced by the local endemic *Doughlatomena splendens* (21.1% of the total individuals) and the plectambonitoid *Metambonites.* Although the sowerbyellids *Gunningblandella* sp. (15.8%) and *Sowerbyella (S.) ampla* (22.1%) are among the three most common species in the former association, in the *Lictorthis* [=*Plectorthis*]-*Metambonites* Association those genera are relatively minor components and only occasionally exceed 20% of the total number of individuals.

Implications for biodiversity

Previous studies on brachiopod biodiversity patterns have mostly been on peri-Iapetus locations (Laurentia, Baltica, and Avalonia), and to a lesser extent degree Mediterranean peri-Gondwana (Stigall *et al.* 2017; Colmenar & Rasmussen 2018; Franeck & Liow 2019; etc.), and therefore they do not fully represent the global patterns. The only exceptions to that are the well-documented Furongian to Mid Ordovician brachiopod faunas of the South China continent (Rong *et al.* 2007; Zhan *et al.* 2011; Zhan & Jin

2014). Unlike the peri-Iapetus area, the first pulses of the brachiopod biodiversification occurred there earlier and in Floian times. South China was the cradle of strophomenoids (Zhan *et al.* 2013) and bryozoans (Xia *et al.* 2007), and both groups played important roles in the Great Ordovician Biodiversification Event (Webby *et al.* 2004) as key components of the Early Palaeozoic Evolutionary Faunas. Thus, there was a diachronous onset in the increase in Ordovician biodiversification across the major early Palaeozoic continents, as noted by Bassett *et al.* (2002).

During the Mid to Late Ordovician, the Chu-Ili Terrane was an integral part of the ensialic volcanic arc separated by the back-arc Zhalair-Naiman basin from the South Kazakh microcontinent. The latter had originated after the Mid Ordovician amalgamation of the Karatau-Naryn and North Tien Shan microcontinental terranes (Popov & Cocks 2017). In the late Darriwilian to Sandbian, Chu-Ili was located in southern tropical latitudes at some distance from the west margin of the Gondwana supercontinent and in relative proximity to the South China Continent, and was a part of the complex system of volcanic island arcs and microcontinents together called the Kazakh Archipelago (Popov & Cocks 2017). Palaeomagnetic data from the Upper Ordovician deposits of North Tien Shan (Bazhenov *et al.* 2003) suggests a position in subequatorial latitudes at 6–9° S. High primary productivity in the surrounding oceans is indicated by the extensive deposition of radiolarian cherts (Tolmacheva *et al.* 2004) which were incorporated within the accretionary wedge in front of the Chu-Ili volcanic arc (Popov *et al.* 2009).

In spite of its relative proximity to South China, the proliferation of the rhynchonelliform brachiopod dominated associations characteristic of the Palaeozoic Evolutionary Faunas was delayed in Chu-Ili until the mid Darriwilian. That mid Darriwilian fauna was documented by Nikitina *et al.* (2006) from numerous sites in the Uzunbulak Formation and lower part of the Berkutsyur Formation (Alakol Limestone) and contains 32 rhynchonelliform genera, which makes its generic richness comparable to that of contemporaneous faunas in the Baltoscandian basin and South China. However, since the mid Darriwilian brachiopods of Chu-Ili have well-defined biofacies differentiation (Fig. 11) and consequently high β-diversity levels, they therefore cannot be considered as a pioneering fauna which had recently invaded the area. Yet the shallow marine Floian to Dapingian limestone units underlying the Uzunbulak Formation contain, in addition to abundant echinoderm ossicles, only gastropods questionably assigned to *Euomphalus* and

unidentified calcareous sponges (Nikitin 1972), but no brachiopods. The Floian to Dapingian brachiopod fauna of the neighbouring North Tien Shan Terrane is dominated by *Tritoechia* and *Archaeorthis* (Rukavishnikova 1956; Popov *et al.* 2001), and is characterised by low diversity but high dominance. There are no precursors of the Darriwilian Chu-Ili faunas anywhere else across the Kazakh terranes.

The late Darriwilian brachiopod fauna of Chu-Ili is newly documented here from the Baigara Formation and contains a total of 16 rhynchonelliform brachiopod genera. There is an apparent decline in richness with time, although that is probably due to sample bias, since late Darriwilian brachiopods are known from a significantly smaller number of localities and from only two transects, and the faunal associations of the outer shelf (BA-4 and BA-5), which contribute significantly to the high diversity of mid-Darriwilian faunas elsewhere, are not represented in Chu-Ili. However, the Margalef's richness scores for the brachiopod associations of BA2 and BA3 show a slight increase in values through the Darriwilian and into Sandbian time, during which origination/extinction rates remained unchanged, whilst the number of originations slightly exceeded the number of extinctions (Fig. 16). Therefore, the observed decrease in faunal turnover through the Darriwilian in Chu-Ili was probably due to normal background origination and extinction rates rather than any significant extinction events.

The core of the Chu-Ili Darriwilian faunas are pantropical genera such as *Aporthophyla*, *Eremotoechia*, *Idiostrophia*, *Leptella*, *Leptellina*, *Taphrodonta*, and *Toquimia*, together with *Neostrophia*, *Trematorthis*, and *Trondorthis* characteristic of the Darriwilian of Laurentia, and *Lepidomena*, which made an earlier appearance in the Darriwilian of the Australian sector of Gondwana (Laurie 1991). Their wide dispersal was probably controlled by the system of equatorial currents. Links with the Darriwilian faunas of South China are accentuated by the presence of early genera of the superfamily Strophomenoidea, as well as by the occurrence of other genera such as *Leptella*, *Martellia*, and *Yangtzeella* (Popov & Cocks 2017).

A significant group of the Kazakh Darriwilian genera were relatively short lived endemics (e.g. *Akadyria*, *Bekella*, *Kujanorthis*, *Oxostrophomena*, *Turanorthis*, and *Uzunbulakia*) and neo-endemics which gave rise to the lineages which spread across the Kazakh terranes and beyond during the Late Ordovician (such as *Acculina*, *Chonetoidea*, *Grammoplecia*, *Katastrophomena*, and *Tenuimena*). Chu-Ili has also preserved within it the earliest known records of the families Bimuriidae (*Asperdelia*),

Paralellelasmatidae (*Kopalina*, *Metacamarella*) and the suborder Orthotetidina (*Kopella*). It is also relevant to note that the general patterns of the depth related brachiopod biofacies developed for the first time during the Darriwilian in Chu-Ili, specifically the lingulide dominated associations of BA1, the rhynchonellide associations of BA2, and the strophomenide dominated associations of BA4-5; all of which persisted without significant changes throughout the late Ordovician. Therefore already by the mid to late Darriwilian, the Chu-Ili volcanic arc was a particular biodiversification centre within the tropical peri-Gondwanan seas.

A total of 68 rhynchonelliform genera are now known from Chu-Ili, a high proportion of the 240 Sandbian genera counted globally by Harper *et al.* (2013); and the generic richness of the Chu-Ili faunas was comparable to that of the entire Laurentian Continent. The beginning of the Sandbian (Time Slice Sa1) was characterised by an increased faunal turnover expressed in almost twice as high origination and extinction rates (Fig. 16) with originations slightly exceeding extinctions. A strong decline in average evenness and increase in dominance values from the late Darriwilian to early Sandbian (Fig. 16) in faunas of BA3 may have been caused by increased environmental stress which triggered the proliferation of new genera with opportunistic life strategies (e.g. *Ishimia*, *Sowerbyella*, and *Chaganella*). That increased taxonomical and ecosystem turnover was perhaps driven locally by a significant reorganisation of habitat caused by the docking of the Mynaral-South Dzhungaria Terrane in front of the Chu-Ili active margin, which resulted in the migration of the magmatic front and consequent formation of the foreland basin (Popov *et al.* 2009; Popov & Cocks 2017). The early Sandbian was also the time of major dispersal of the brachiopod faunas across the whole of the Kazakh Archipelago, with Chu-Ili as the most likely centre of origin of newly settled faunas. However, brachiopod faunas in the newly populated areas elsewhere in the Kazakh terranes did not reach the α- and β-biodiversity levels observed in Chu-Ili, apart from in the Chngiz-Tarbagatai island arc system (Popov & Cocks 2014; 2017).

In the late Sandbian (Sa2) the brachiopod extinction rates in Chu-Ili declined significantly to become the lowest observed through the Mid and Late Ordovician (Fig. 16). It was also the acme in the biodiversification of the Chu-Ili brachiopod faunas, when its richness increased from 39 to 53 genera. The most significant rise in biodiversity from the early to late Sandbian occurred in the brachiopod associations of BA3, with Margalef's richness index climbing from 1.816 to 3.334 in average value

(Fig. 16). That was mainly because of the proliferation of the medium to high rich *Parastrophina-Kellerella* and *Acculina-Dulankarella* associations which were closely linked to algal and microbial carbonate build-ups (Popov *et al.* 2002). In the early Sandbian, *Acculina* and *Dulankarella* were common components within the BA3 *Acculina-Sowerbyella* Association, but after the late Sandbian both genera migrated to environments closely associated with carbonate build-ups, while *Sowerbyella* retained its dominance on the shallow clastic shelves of BA2-3 until the end of the Katian (Popov & Cocks 2017). Apart from the obvious taxonomic links of the strophomenide and orthide components of the late Sandbian brachiopod faunas associated with carbonate build-ups and the early Sandbian faunas of BA2-3, expressed in the occurrence of such genera as *Acculina*, *Bandaleta*, *Bellimurina*, *Christiania*, *Dulankarella*, *Grammoplecia*, *Leptaena* (*Ygdrasilomena*), *Ptychoglyptus*, and *Sowerbyella*, they are quite distant from all the other Sandbian brachiopod faunas. In the Principal Components Analysis the *Parastrophina-Kellerella* and *Acculina-Dulankarella* associations show positive scores along the first maximum direction variation (Fig. 13) unlike other Sandbian associations, while in the Principal Coordinate Analysis (Morisita Similarity) only these two associations are characterised by high negative scores along the second maximum direction variation (Fig. 14). In the Cluster Analysis (Morisita Similarity) the *Parastrophina-Kellerella* and *Acculina-Dulankarella* associations are very weakly linked within a single third order cluster with the early Sandbian *Eoanastrophia* and *Chaganella* associations (Fig. 12), which were possible precursors of the late Sandbian brachiopod associations which inhabited carbonate build-ups.

The pentamerides of the superfamily Camerelloidea (*Didymelasma*, *Eoanastrophia*, *Ilistrophina*, *Liostrophia*, *Parastrophina*, *Plectosyntrophia*?, and *Schizostrophina*), and the early spirebearers of the superfamilies Protozygoidea (*Costistriispira*, *Kellerella*, *Nikolaispira*, and *Rozmanospira*) and Atrypoidea (*Pectenospira*) became increasingly abundant through the Sandbian in the brachiopod associations from algal limestones with dasyclad algae and carbonate build-ups BA3). The local stratigraphical range of *Eoanastrophia* in Chu-Ili is restricted to the early Sandbian; but in the Katian it became relatively common across Mediterranean peri-Gondwana as a component of the *Nicolella* Community (Havlíček 1981; Colmenar & Rasmussen 2018). *Liostrophia* and *Parastrophina* made later appearances in Laurentia, while *Plectosyntrophia*? and *Schizostrophina* are known from the slightly older or contemporaneous

Ordovician deposits of North China (Rong *et al.* 2017). The earliest spire-bearing brachiopods of the superfamily Protozygoidea first appear approximately synchronously in the early Sandbian of Laurentia and Chu-Ili (Popov *et al.* 1999; Copper 2002). The earliest and most primitive protozygoid atrypide in Chu-Ili is *Rozmanospira*, which is characterized by simple brachial supports with a single incomplete whorl and has no jugum. The late Sandbian protozygoid genera *Kellerella* and *Nikolaispira* had more complex spiralia with laterally directed cones of up to five whorls, which makes them unique among the early atrypides. The earliest atrypidine *Pectenospira* had simple centrodorsal spiralia with fewer than two whorls. A significant radiation of the suborder occurred in Chu-Ili in early to mid Katian time, when the atrypidines spread widely across the shallow self and acquired complex brachial supports with dorsomedially directed spiralial cones (Popov *et al.* 1999). Remarkably, in spite of the observed disparity in the morphology of the brachial supports, there are no taxa with a complete jugum known among the Kazakh Ordovician atrypides, suggesting their early divergence and separation from the Laurentian Ordovician atrypide lineages, which are almost invariably characterised by a complete jugum (Copper 2002). That further confirms that Chu-Ili was the most important global centre of origin and early diversification of brachiopods with calcified brachial supports.

The late Sandbian in Chu-Ili showed a temporary decline of the rhynchonellide associations of BA2, which were replaced by the *Testaprica* and *Mabella-Sowerbyella* associations (Figs 11–14), with plectambonitoids as the dominant taxa. However, the rhynchonellides returned to their dominance on the shallow marine clastic shelves (BA2) of the Kazakh microcontinents and island arcs after the early Katian, mainly due to the proliferation of such invasive genera as *Lydirhyncha* and *Rhynchotrema* (Popov & Cocks 2017). There was a sharp drop in generic richness of Chu-Ili rhynchonelliform brachiopod faunas from 53 to 34 in the early Katian (Ka1), with extinction rates exceeding origination rates for the first time (Popov & Cocks 2017, fig. 14). However, that is probably at least partly due to sampling bias, because carbonate build-ups of that age are confined mainly to the poorly accessible area of the northern Betpak-Dala desert today, where the related brachiopod fauna has only been documented from a single locality (Nikitin & Popov 1996; Nikitin *et al.* 1996). Yet the major impact on the early Katian brachiopod diversity was from significant decline of the brachiopod faunas in the deep shelf settings

(BA4-5), also recognised by Popov & Cocks (2017) in other parts of Kazakhstan.

The general biodiversity trend of the Darriwilian to early Katian rhynchonelliform brachiopod faunas from Chu-Ili with its acme in the late Sandbian (Sa1) and decline in the early Katian (Ka1) is broadly similar to that depicted for the genus-level diversity trend of occurrence data by Kröger & Lintulaakso (2017), but differs from the global Ordovician generic brachiopod diversity trend shown by Harper *et al.* (2013) and Servais & Harper (2018, fig. 4) in which the biodiversity maximum was delayed until the late Sandbian, while extinction rates were down to the minimum and exceeded origination rates not in the late Sandbian, but in the early Katian. That was probably a local phenomenon due to extensive growth of the late Sandbian microbial/algal carbonate build-ups and proliferation of associated reef-dwellers (Nikitin *et al.* 1974; Popov *et al.* 2002).

During the Early to Mid Ordovician, the peri-Gondwanan South China continent was the most important centre of the origin and dispersal of several groups within the Palaeozoic Evolutionary Faunas, including rhynchonelliform brachiopods (Rong *et al.* 2017; Zhan *et al.* 2011; Zhan & Jin 2014). In that time a new brachiopod biodiversity 'hot spot' arose in tropical latitudes off the west Gondwana coast, in the area occupied by the Kazakh Archipelago, including the Chu-Ili island arc. Thus Chu-Ili was the cradle for the initial evolution and dispersal of such brachiopod groups as the suborder Atrypidina and superfamily Orthotetoidea, both of which survived the Ordovician terminal extinction and played a prominent role in the Silurian and Devonian faunas. It was also the centre of origin of the camerelloid family Parallelasmatidae, which, although only a minor component of the Ordovician faunas, is considered by some workers (e.g. Sapelnikov 1985) to have been the ancestral group for the suborder Pentameridina. The early pentameridines of the superfamily Camerelloidea flourished later in Chu-Ili during the late Katian (Sapelnikov & Rukavishnikova 1975). The rich rhynchonelliform brachiopod associations from deep-water settings (BA4-5) are known in Chu-Ili from the Darriwilian, earlier than in other parts of Gondwana and its satellites (the Kazakh terranes and the South and North China Microcontinents), although there is the earliest documented evidence of the migration of rhynchonelliform brachiopods (*Akadyria*) down into bathyal depths. During the Sandbian, a strong biodiversification pulse was triggered in Chu-Ili by the growth of carbonate build-ups, which were a prelude to the reef communities associated with microbial, algal, and metazoan build-ups across the Kazakh Archipelago and beyond.

Localities

There follows a list of fossil localities, together with some of their significant brachiopods and numbers of complete shells, ventral valves, and dorsal valves (in brackets after each taxon).

North Betpak-Dala

Locality 154. – Tastau Formation (late Darriwilian); Karakan Ridge [46°31′N; 70°20′E], graded siltstones and siliceous argillites with *Akadyria simplex* Nikitina *et al.*, 2006 and *Oxolosia*? sp., the trilobite *Pricyclopyge* sp. and graptolites *Expansograptus* sp., *Isograptus* sp., *Loganograptus* sp., *Trigonograptus praelongus* Keller *in* Keller & Lisogor, 1954, *Phyllograptus* sp. (collected in 1974 by L. E. Popov. and D. T. Tsai).

Locality 162. – Tastau Formation (late Darriwilian); Golubaya Gryada [46°27′50″N, 70°25′20″]; graded siltstones and siliceous argillites (late Darriwilian) with *Broeggeria* cf. *B. putilla* Tenjakova, 1989 (0:1:0) (collected in 1974 by L. E. Popov. and D. T. Tsai).

Locality 163. – Tastau Formation (late Darriwilian); Golubaya Gryada [46°25′14″N, 70°26′22″E]; graded siltstones and siliceous argillites (Darriwilian) with *Broeggeria* cf. *B. putilla* Tenjakova, 1989 (0:0:1) and graptolites including *Eoglyptograptus dentatus* (Brongniart, 1828) (collected in 1974 by L. E. Popov. and D. T. Tsai).

Locality 166. – Takyrsu Formation (late Darriwilian to early Sandbian); Golubaya Gryada [46°28′46″N; 70°26′12″E]; Kypchak Limestone (late Darriwilian–early Sandbian) brownish-red bedded limestone with trilobites and *Altynorthis tabylgatensis* n. sp. (*c.* 50 disarticulated dorsal and ventral valves and their fragments), *Buminomena* sp. (0:0:1), *Eridorthis* sp. (0:2:0), *Esilia* (0:1:0), Hesperorthidae gen. et sp. indet. (0:4:5), *Mabella* sp. (1:0:0), *Ptychoglyptus* sp. (0:2:0), *Shlyginia* sp. (0:1:0); *Sowerbyella (S.) verecunda baigarensis* n. subsp. (0:5:1) (collected in 1974 by L. E. Popov. and D. T. Tsai).

South Betpak-Dala

Locality 625/1. – Baigara Formation (late Darriwilian); area *c.* 6 km south-west of Baigara Mountain

(precise position unknown); *Bimuria* sp. (1:0:0), *Lepidomena betpakdalensis* n. sp. (6:0:0) (collected by T. B. Rukavishnikova).

Locality 625/3. – Baigara Formation (late Darriwilian); area *c.* 6 km south west of Baigara Mountain (precise position unknown) *Colaptomena insolita* (Nikitina, 1985) (4:0:0), *Katastrophomena rukavishnikovae* (Nikitina, 1985) (1:0:0), *Lepidomena betpakdalensis* n. sp. (16:0:0), *Sonculina baigarensis* n. sp. (6:0:0) (collected by T. B. Rukavishnikova).

Locality 765-e. – Baigara Formation (late Darriwilian); area *c.* 6 km south west of Baigara Mountain (precise position unknown); *Acculina acculica* Misius in Misius & Ushatinskaya, 1977 (6:0:0), *Plectocamara* sp. (2:0:0) (collected by T. B. Rukavishnikova).

Area 10 km south-west of Lake Alakol

Locality 735. – [44°49′24″N; 74°6′59″E] Kopkurgan Formation (middle Katian); dark grey fine-grained sandstone with shell beds; *Acculina*? sp. (1:3:3), *Anoptambonites kovalevskii* Popov, Nikitin & Cocks, 2000 (0:6:6), *Bokotorthis kasachstanica* (Rukavishnikova, 1956) (0:2:4), *Christiania proclivis* Popov & Cocks, 2006 (1:4:6), *Doughlatomena splendens* n. sp. (2:4:16), *Grammoplecia* sp. (0:0:5), *Gunningblandella* sp. (1:10:5), *Katastrophomena* sp. (0:1:0), *Leangella (Leangella)* sp. (0:8:3), *Lictorthis* sp. (0:1:0), *Lydirhyncha tarimensis* (Sproat & Zhan, 2018) (0:0:2), *Onniella* sp. (0:1:1), *Petrocrania* sp. (0:0:2), *Phaceloorthis*? *corrugata* n. sp. (0:6:3), *Rhipidomena* sp. (0:1:0), *Sowerbyella (Sowerbyella) ampla* (Nikitin & Popov, 1996) (3:14:18) (collected in 1993 by L. E. Popov).

Area 8 km south-west of Lake Alakol

Locality 727. – [44°48′35″N; 74°2′1″E] Kopkurgan Formation (middle Katian); dark grey fine-grained sandstone with shell beds; *Anoptambonites* sp. (0:1:0), *Parastrophina* sp. (0:0:1) (collected in 1993 by L. E. Popov).

Area 7 km south-west of Lake Alakol

Locality 8120. – [44°48′59″N, 74°01′47″E] Berkutsyur Formation (Sandbian); light grey nodular limestone with dasyclad algae, *Altynorthis vinogradovae* n. sp. (0:1:0), *Ishimia* aff. *I. ishimensis* Nikitin, 1974 (3:2:2), *Bandaleta* sp. (0:2:0) (collected in 1981 by I. F. Nikitin & L. E. Popov).

Locality 8120/4. – [44°48′59″N, 74°01′47″E] Berkutsyur Formation (Sandbian); light grey nodular limestone with dasyclad algae, *Eoanastrophia kurdaica* Sapelnikov & Rukavishnikova, 1975 (1:0:0), *Eridorthis* sp. (0:0:1), *Ishimia* aff. *I. ishimensis* Nikitin, 1974 (1:2:1) (collected in 1981 by I. F. Nikitin & L. E. Popov).

Locality 8121. – [44°48′58″ N; 74° 2′13″ E] Berkutsyur Formation (Sandbian); light-grey bioclastic limestone with *Baitalorhyncha rectimarginata* n. sp. (1:0:0); *Cooperea* sp. (0:1:1), *Esilia* cf. *E. tchetverikovae* (0:0:1), *Eichwaldia* sp. (0:1:0), *Eoanastrophia kurdaica* Sapelnikov & Rukavishnikova, 1975 (1:0:0), *Eridorthis* sp. (0:0:1), *Ilistrophina tesikensis* Popov *et al.*, 2002 (0:1:1), *Paraoligorhyncha?* sp. (1:0:0), *Ptychoglyptus* sp. (0:1:0), *Rozmanospira mica* (Nikitin & Popov, 1984) (11:0:0), *Sowerbyella (S.)* cf. *S. acculica* (0:0:1), *Synambonites* sp. (0:0:1), *Triplesia* sp. (1:0:0) (collected in 1981 by I. F. Nikitin & L. E. Popov).

Locality 8122. – [44°48′55″N, 74°01′58″E] Kopkurgan Formation (Sandbian); black graptolitic shale with *Aploobolus tenuis* n. sp. (1:4:4), *Chonetoidea?* sp. (0:1:0), *Costistriispira?* sp. (0:1:0), *Tenuimena* aff. *T. planissima* Nikitina *et al.*, 2006 (4:3:6), *Elliptoglossa magna* Popov, 1977 (3:[17]) (collected in 1981 by I. F. Nikitin & L. E. Popov).

Locality 8124. – [44°48′58″N, 74°01′56″E] Berkutsyur Formation (Sandbian); light-grey bioclastic limestone with dasyclad algae, *Altynorthis vinogradovae* n. sp. (0:1:3), *Bandaleta* sp. (0:0:3), *Eoanastrophia kurdaica* Sapelnikov & Rukavishnikova, 1975 (4:0:0), *Eridorthis* sp. (3:0:0); *Ilistrophina tesikensis* Popov *et al.*, 2002 (1:0:0), *Ishimia* aff. *I. ishimensis* Nikitin, 1974 (9:5:2), *Leptaena (Ygdrasilomena)* sp. (0:1:0) (collected in 1981 by I. F. Nikitin & L. E. Popov).

Locality 8125. – [44°48′54″N, 44°48′54″N] Berkutsyur Formation (Sandbian); light yellowish-brown calcareous siltstone with *Altynorthis vinogradovae* n. sp. (0:2:1) (collected in 1981 by I. F. Nikitin & L. E. Popov).

Locality 8233. – [44°48′54″N, 74°02′43″E] Berkutsyur Formation (Sandbian); light-grey bioclastic limestone with dasyclad algae, *Altynorthis vinogradovae* n. sp. (0:1:0), *Bandaleta* sp. (3:2:3), *Eoanastrophia kurdaica* Sapelnikov & Rukavishnikova, 1975 (7:1:1), *Eridorthis* sp. (6:1:0), *Esilia* cf. *E. tchetverikovae* Popov & Nikitin, 1985 (1:0:1) (collected in 1982 by I. F. Nikitin & L. E. Popov).

Locality 8235. – [44°48′54″N, 74°01′44″E] Kopkurgan Formation (late Sandbian); argillaceous matrix in olistostrome horizon with *Kassinella simorini* n. sp. (1:1:3), *Pionodema opima* Popov et al., 2002 (0:2.2), *Tesikella necopina* (Popov, 1980b) (0:2:1), *Shlyginia* sp. (0:1:1) (collected in 1982 by I. F. Nikitin & L. E. Popov).

Locality 8236a. – [44°48′51″N, 74°02′02″E] Berkutsyur Formation (Sandbian); light-grey bioclastic limestone with dasyclad algae, *Bandaleta* sp. (0:1:0), *Eridorthis* sp. (2:0:0); *Esilia* cf. *E. tchetverikovae* (Nikitin & Popov, 1985) (0:0:1), *Ishimia* aff. *I. ishimensis* Nikitin, 1974 (0:1:0) (collected in 1982 by I. F. Nikitin & L. E. Popov).

Locality 8236. – [44°48′51″N, 74°02′02″E] Kopkurgan Formation (Sandbian); black graptolitic shale, Kopkurgan Formation (Sandbian) with *Chonetoidea?* sp. (2:0:2). *Elliptoglossa magna* Popov, 1977 (0:0:1) (collected in 1982 by I. F. Nikitin & L. E. Popov).

Area 5.5 km south-west of Lake Alakol

Locality N-12. – [approximate coordinates 44°49′N; 74.04′E] Berkutsyur Formation (Darriwilian), brachiopod shell bed in horizontal and cross laminated arkosic sandstones with *Ancistrorhyncha modesta* Popov in Nikiforova & Popov, 1981 (0:7:9) (collected in 1981 by I. F. Nikitin)

Area 4 km south-west of Lake Alakol

Locality 812. – [44°49′24.48″N; 74° 6′59.50″E] Berkutsyur Formation; light yellowish-brown calcareous siltstone with *Acculina acculica* Misius in Misius & Ushatinskaya, 1977 (0:4:5), *Altynorthis vinogradovae* n. sp. (1:2:1), *Buminomena* sp., (0:1:1), *Christiania* cf. *C. tortuosa* Popov, 1980a (0:3:2), *Ishimia* aff. *I. ishimensis* Nikitin, 1974 (3:1:5), *Isophragma princeps* Popov, 1980a (1:3:3), *Mabella* sp. (0:5:4), *Sowerbyella (S.) verecunda baigarensis* n. subsp. (2:10:6), *Testaprica alperovichi* n. sp. (2:2:3) (collected in 1981 by I. F. Nikitin & L. E.Popov).

Locality 813. – [44°49′40″N; 74° 6′38″E] Berkutsyur Formation; light yellowish-brown calcareous siltstone with *Altynorthis vinogradovae* n. sp. (3:8:17), *Buminomena* sp., (0:0:3), *Dulankarella larga* Popov et al., 2002 (0:1:1), *Ishimia* aff. *I. ishimensis* Nikitin, 1974 (11:5:5), *Isophragma princeps* Popov, 1980a (1:3:3), *Leptaena (Ygdrasilomena)* sp. (0:1:0); *Sowerbyella (S.) verecunda baigarensis* n. sp. (2:10:6), *Testaprica alperovichi* n. sp. (2:14:9) (collected in 1981 by I. F. Nikitin & L. E. Popov).

Locality 814. – [44°49′35″N; 74° 6′47″E] Berkutsyur Formation; light yellowish-brown calcareous siltstone with *Altynorthis vinogradovae* n. sp. (4:7:4), *Testaprica alperovichi* n. sp. (2:8:3), *Ishimia* aff. *I. ishimensis* Nikitin, 1974 (0:1:0) (collected in 1981 by I. F. Nikitin & L. E. Popov).

Locality 815. – [44°49′35″N: 74°06′06″E] Berkutsyur Formation; light yellowish-brown calcareous siltstone with *Acculina acculica* Misius in Misius & Ushatinskaya, 1977 (3:0:0), *Altynorthis vinogradovae* n. sp. (3:0:1), *Buminomena* sp., (3:0:0), *Christiania tortuosa* Popov, 1980a (0:0:1), *Dulankarella larga* Popov *et al.*, 2002 (0:1:2); *Ishimia* aff. *I. ishimensis* Nikitin, 1974 (4:3:3), *Isophragma princeps* Popov, 1980a (0:0:1), *Sowerbyella (S.) acculica* Misius *in* Misius & Ushatinskaya, 1977 (0:2:1); *Mabella* sp. (0:1:0) (collected in 1981 by I. F. Nikitin & L. E. Popov).

Locality 816. – Kopkurgan Formation; olistolith of light yellowish-brown silty bioclastic limestone with *Acculina acculica* Misius in Misius & Ushatinskaya, 1977 (1:3:3), *Altynorthis vinogradovae* n. sp. (1:0:0), *Bandaleta* sp. (1:6:3), *Bimuria karatalensis* n. sp. (1:0:0), *Buminomena* sp., (0:0:1), *Dulankarella larga* Popov *et al.*, 2002 (4:5:6), *Esilia* cf. *E. tchetverikovae* Nikitin & Popov, 1985 (0:0:1), *Sowerbyella (S.)* cf. *S. acculica* (0:1:0), *Sowerbyella (S.) verecunda baigarensis* n. subsp. (0:4:4) (collected in 1981 by I. F. Nikitin & L. E. Popov).

Locality 817. – [44°49′28″N, 74° 6′7″E] Berkutsyur Formation; light yellowish-brown calcareous siltstone with *Accilina acculica* Misius in Misius & Ushatinskaya, 1977 (2:0:0), *Dolerorthis expressa* Popov, 1980b (0:0:1), *Dulankarella larga* Popov *et al.*, 2002 (0:3:3), *Pionodema opima* Popov *et al.*, 2002 (2:0:0), *Sowerbyella (S.) verecunda baigarensis* n. subsp. (2:14:4) (collected in 1981 by I. F. Nikitin & L. E. Popov).

Locality 817a. – [44°49′28″N, 74° 6′6″E] Berkutsyur Formation (Sandbian); greenish grey fine grained sandstone and siltstone intercalations, with *Ectenoglossa sorbulakensis* Popov, 1980b (0:4:4) (collected in 1981 by I. F. Nikitin & L. E. Popov).

Locality 8234. – [44°49′28″N, 74° 6′5″E] Berkutsyur Formation (Sandbian); brownish-grey calcareous sandstone with brachiopod coquinas, *Altynorthis tabylgatensis* (Misius, 1986) (1:0:0), *Baitalorhyncha rectimarginata* n. sp. (0:26:26), *Costistriispira proavia* n. sp. (0:13:25), *Ectenoglossa sorbulakensis* Popov, 1980b (0:1:0), *Trematis?* sp. (0:0:1) (collected in 1982 by I. F. Nikitin & L. E. Popov).

Area 6 km south-west of Baigara Mountain

Locality 1020. – [45°11′20″N; 72°21′1″ E] Baigara Formation (late Darriwilian – early Sandbian); dark, greyish-brown calcareous sandstone with brachiopod and bivalve coquinas; *Altynorthis betpakdalensis* n. sp. (1:0:0), *Ancistrorhyncha modesta* Popov in Nikiforova & Popov, 1981 (1:43:43), *Katastrophomena rukavishnikovae* (Nikitina, 1985) (5:0:0), *Sowerbyella (S.) verecunda baigarensis* n. subsp. (1:2:2) (collected in 1974 by L. E.Popov & D. T. Tsai).

Locality 1021. – [45°11′20″ N; 72°21′1″ E] Baigara Formation (late Darriwilian – early Sandbian); light grey, silty limestone with *Acculina acculica* Misius in Misius & Ushatinskaya, 1977 (31:1:1), *Altynorthis betpakdalensis* n. sp. (60:0:1), *Bimuria* sp. (3:2:0), *Colaptomena insolita* (Nikitina, 1985) (22:0:1), *Ectenoglossa sorbulakensis* Popov, 1980b (0:1:0), *Katastrophomena rukavishnikovae* (Nikitina, 1985) (25:1:0), *Leptellina* sp. (58:0:2), *Plectocamara* sp. (1:0:0), *Scaphorthis recurva* Nikitina, 1985 (21:1:0), *Sonculina baigarensis* n. sp. (6:1:0), *Sowerbyella (S.) verecunda baigarensis* n. subsp. (51:6:4) (collected in 1974 by L. E.Popov & D. T. Tsai).

Locality 1022. – [45°11′20″N; 72°21′1″E] Baigara Formation (late Darriwilian – early Sandbian); light grey, nodular limestone with abundant dasyclad algae and brachiopods *Acculina acculica* Misius *in* Misius & Ushatinskaya, 1977 (4:0:0), *Altynorthis betpakdalensis* n. sp. (239:0:0), *Apatomorpha akbakaiensis* n. sp. (9:0:0), *Bimuria* sp. (10:0:0), *Colaptomena insolita* (Nikitina, 1985) (57:0:0), *Eremotoechia inchoata* Popov *et al.*, 2000 (6:0:0), Furcitellinae gen. et sp. indet. (3:0:0), *Grammoplecia* aff. *G. globosa* (Nikitin & Popov, 1985) (9:0:0), *Katastrophomena rukavishnikovae* (Nikitina, 1985) (23:0:0), *Lepidomena betpakdalensis* n. sp. (52:0:0), *Leptellina* sp. (8:0:0), *Multispinula* sp. (0:1:0), *Scaphorthis recurva* Nikitina, 1985 (15:0:0), *Sonculina baigarensis* n. sp. (26:0:0), *Sowerbyella (S.) verecunda baigarensis* n. subsp. (38:0:0), *Trematis* aff. *T. parva* Cooper, 1956 (1:0:0) (collected in 1974 by L. E. Popov & D. T. Tsai).

Locality 1023. – [45°11′20″N; 72°21′1″E] Baigara Formation (late Darriwilian – early Sandbian); light grey, nodular limestone with abundant dasyclad algae and *Altynorthis betpakdalensis* n. sp. (153:0:0), *Apatomorpha akbakaiensis* n. sp. (20:0:0), *Bimuria* sp. (4:0:0), *Colaptomena insolita* (Nikitina, 1985) (12:0:0), *Eremotoechia inchoata* Popov *et al.*, 2002

(32:0:0), *Grammoplecia* aff. *globosa* (Nikitin & Popov, 1985) (9:0:0), *Katastrophomena rukavishniko-vae* (Nikitina, 1985) (8:0:0), *Lepidomena betpak-dalensis* n. sp. (42:1:0), *Leptellina* sp. (1:0:0), *Sonculina baigarensis* n. sp. (12:0:0), *Sowerbyella (S.) verecunda baigarensis* n. subsp. (25:0:0) (collected in 1974 by L. E.Popov & D. T. Tsai).

West side of the River Karatal, Baigara

Locality 1025. – [45°15′45″N; 72°5′52″E] Baigara Formation (late Darriwilian – early Sandbian); light grey, nodular limestone with *Altynorthis betpakdalen-sis* n. sp. (6:0:0), *Eremotoechia inchoata* Popov *et al.*, 2000 (2:0:0), *Sonculina baigarensis* n. sp. (2:0:0), *Sowerbyella (S.) verecunda baigarensis* n. subsp. (4:0:0) (collected in 1974 by L. E. Popov & D. T. Tsai).

Locality 1026. – [45°15′48″N; 72°5′59″E] Baigara Formation (late Darriwilian – early Sandbian), dark grey siltstone with *Altynorthis tabylgatensis* n. sp. (0:3:6), *Bimuria karatalensis* n. sp. (0:7:7),. *Christiania* cf. *C. tortuosa* Popov, 1980a (0:5:3), *Eremotoechia spissa* (0:0:1), *Grammoplecia globosa* (Nikitin & Popov, 1985) (0:26:24), *Infurca* sp. (0:1:3), *Kajnaria derupta* Nikitin & Popov, 1984 (1:4:0), *Pseudocrania karatalensis* Popov *in* Nazarov & Popov, 1980 (0:0:5), *Sowerbyella (S.) verecunda baigarensis* n. subsp. (0:6:12); *Titanambonites* cf. *T. magnus* Nikitin, 1974 (0:1:0) (collected in 1974 by L. E. Popov & D. T. Tsai).

Locality 1026a. – [45°15′59″N; 72°5′59″E] Baigara Formation (late Darriwilian – early Sandbian) with *Bimuria karatalensis* (0:1:0) (collected in 1974 by L. E. Popov & D. T. Tsai).

Locality 1026b. – [45°15′46″N; 72°6′4″E] Baigara Formation (late Darriwilian – early Sandbian), light grey, nodular highly argillaceous limestone with *Altynorthis tabylgatensis* (Misius, 1986) (7:0:1), *Atele-lasma* sp. (2:0:0), *Bimuria karatalensis* n. sp. (26:14:9), *Christiania* cf. *C. tortuosa* (11:0:1), *Eremo-toechia spissa* Popov *et al.*, 2000 (20:0:0), *Grammo-plecia globosa* (Nikitin & Popov, 1985) (67:0:0), *Kajnaria derupta* Nikitin & Popov, 1984 (9:0:0), *Son-culina* cf. *S. prima* (16:0:0), *Pseudocrania karatalensis* Popov *in* Nazarov et Popov, 1980 (0:1:0), *Sowerbyella (S.) verecunda baigarensis* n. subsp. (2:0:0), *Titanam-bonites* cf. *T. magnus* Nikitin, 1974 (2:0:0), *Trematis* aff. *T. parva* Cooper, 1956 (0:0:1) (collected in 1974 by L. E. Popov & D. T. Tsai).

Locality 1028. – Baigara Formation (late Darriwil-ian-early Sandbian) at the east side of Karatal River [45°15′24″N; 72°6′31″E]; greenish grey siltstone with

Altynorthis tabylgatensis (Misius, 1986) (0:11:9), *Grammoplecia globosa* (Nikitin & Popov, 1985) (0:1:2), *Sowerbyella (S.) verecunda baigarensis* n. subsp. (0:4:2) (collected in 1974 by L. E. Popov & D. T. Tsai).

Other areas

Locality 388. – Berkutsyur Formation (Sandbian), at the area 15 km west of Chimpek Bay [precise coordi-nates unavailable], West Balkhash Region; dark grey bedded limestone with *Chaganella chaganensis* Nikitin, 1974 (0:6:8) and *Eridorthis* sp. (0:2:0) (collected in 1976 by L. E. Popov).

Locality 388a. – Kopkurgan Formation (Sandbian) at the area 15 km west of Chimpek Bay [precise coordi-nates unavailable], West Balkhash Region; black argillite with lingulide *Meristopacha* sp. (0:7:7) (col-lected in 1976 by L. E. Popov).

Locality 538. – Anderken Formation (Sandbian) at the Burultas site [45° 5′1″N; 73°29′19″E] (see Popov *et al.* 2002), dark gray argillite with yellowish-brown calcareous nodules, with trilobites and a few bra-chiopods, including *Kassinella simorini* n. sp. (0:3:0) and *Gunningblandella* sp. 2 (0:1:0) (collected in 1993 by L. E. Popov).

Locality 1501. – Unnamed Formation (Katian), area in vicinity of Ergibulak well, Ergenekty Mountains [approximate coordinates 46°25′N, 71°1′E], north-ern Betpak-Dala; greenish grey siltstones with *Boko-torthis kasachstanica* (0:1:0), *Qilianotryma* cf. *Q. suspectum* Popov *in* Nikiforova *et al.*, 1982 (0:0:1), *Shlyginia extraordinaria* (Rukavishnikova, 1956) (0:9:1), *Sowerbyella (S.) ampla* (Nikitin & Popov, 1996) (0:0:2) (Collected in 1981 by I. F. Nikitin, D. T. Tsai & L. E. Popov).

Locality 8126. – [44°49′26″N 74°03′56″E] unnamed formation (late Katian); Peninsula on southern cost of unnamed salt lake at 7 km south west of Lake Ala-kol, bioclastic limestone in brownish-red fine to medium grained sandstones with bidirectional cross-lamination with *Anoptambonites* sp. (0:3:1) (col-lected in 1981 by I. F. Nikitin & L. E. Popov).

Locality 11221. – Rgaity Formation (Darriwilian) near the Talapty winter hut [at about 43°11′N 74°46′S], Rgaity river basin, South Kendyktas Range, North Tien Shan Terrane, tuffaceous fine grained sandstones and siltstones with *Acculina acculica* Misius *in* Misius & Ushatinskaya, 1977 (0:3:2), *Colaptomena insolita* (0:5:7), *Paralenorthis rgaitensis* (Nikitina, 1985) (0:5:1), *Scaphorthis recurva*

Nikitina, 1985 (0:13:15) (collected in 2006 by A. V. Mikolaichuk).

Systematic palaeontology

Remarks. – Dimensions are given in mm in brackets after some specimen numbers: W, L, T, maximum width, length, thickness of the shell; Lv, Ld, maximum length of ventral and dorsal valve; L/W, length/width ratio; Iw, Il maximum width and length of interarea; Bl, Bw, brachiophore length and width; Ml, Mw length and width of muscle field; Sl, median septum length; X, mean; S, standard deviation from the mean; OR, observed range; max, maximum value; min, minimum value; N, number of measured or counted specimens. References to family group taxa and above are not given, since none are newly erected here apart from the Family Kellerellidae, and all may be found in the *Treatise on invertebrate paleontology* (Kaesler 1997–2004).

Repositories. – The specimens are in the National Museum of Wales, Cardiff (NMW), The Natural History Museum (NHM, UK), London (BC), the Central Geological Research and Exploration Museum, St Petersburg, Russia (CNIGR), and the Institute of Geological Sciences, Almaty, Kazakhstan (IGNA).

Class Lingulata Goryanski & Popov, 1985

Order Lingulida Waagen, 1885

Superfamily Linguloidea Menke, 1828

Family Obolidae King, 1846

Subfamily Obolinae King, 1846

Genus *Aploobolus* n. gen., n. sp.

Derivation of name. – After Greek, *aplós*, simple and *obolós*, silver coin.

Type species and only species. – *Aploobolus tenuis* n. gen et n. sp., from the Kopkurgan Formation (Upper Ordovician, Sandbian), of West Balkhash Region, South Kazakhstan.

Diagnosis. – Shell subcircular, subequally biconvex, with well-developed pseudointerareas and flexure lines on propareas of both valves, ornamented with dence faint, ridge-like concentric rugellae, lacking dorsal median ridge. Ventral interior weakly impressed, except for the paired ventral umbonal

muscle scars divided by indistinct pedicle nerve impression. No dorsal median ridge.

Remarks. – In spite of a relatively simple shell morphology and superficial similarity to such obolide genera as *Apatobolus* Popov in Nazarov & Popov, 1980, *Divobolus* Sutton in Sutton *et al.*, 1999, and *Paldiskites* Havlíček, 1982, as revised by Mergl, 2002, *Aploobolus* is different in its unique combination of morphological features. *Paldiskites* was synonymised with *Apatobolus* by Holmer & Popov, but subsequent revision by Mergl (2002) confirmed that it is a valid obolid genus. It is morphologically closest to *Aploobolus*; however, it can be distinguished from the latter in having dorsal propareas divided by flexure lines, a concentric ornament of dense rugellae, and in the complete absence of radial ornamentation. *Aploobolus* obviously differs from *Apatobolus* in having well developed pseudointerareas with flexure lines on the propareas of both valves and in the absence of paired ridges dividing the ventral umbonal muscle scars. Unlike *Divobolus*, *Aploobolus* has well defined flexure lines on the propareas of both valves, it completely lacks a dorsal median ridge, and is characterised by very poorly impressed features inside both valves.

Aploobolus tenuis n. sp.

Plate 1, figures 2–11, 15–16

Derivation of name. – After Latin, *tenuis* – slender, faint.

Holotype. – NMW 98.28G.2135, ventral internal mould (Pl. 1, figs 2, 5; Lv = 8.9, W = 7.8, Iw = 5.0, Il = 1.6) from the Kopkurgan Formation (Sandbian), Locality 8122, 6 km south-west of Lake Alakol, West Balkhash Region.

Paratypes. – Locality 8122: one disarticulated shell internal and external moulds, including NMW 98.28G.2134.1-2 (Lv, 3.5, Ld, 3.5, W, 3.8; Pl. 1, figs 7, 8); four ventral valves, including NMW 98.28G.2141–2143 (Lv, 4.7, W, 4.3; Pl. 1, fig. 10), 2144 (Lv, 5.1, W, 5.1; Pl. 1, fig. 9); four dorsal valves, including NMW 98.28G. 2136, (Ld, 10.6; W, 10.5, Il, 0.8, Iw, 3.6; Pl. 1, figs 3, 6), 2137, 2138 (Ld, 4.7, W, 5.4; Pl. 1, fig. 16), 2139, (Ld, 5.1, W, 5.8; Pl. 1, fig. 10), 2140, dorsal valve (Ld, 3.8; W, 3.8; Pl. 1, fig. 15).

Diagnosis. – As for the genus.

Description. – Shell subequally biconvex, almost subcircular, about 104% as long as wide with maximum width at mid-length. Anterior commissure broadly rounded, rectimarginate. Ventral valve gently convex with a pointed umbo. Pseudointerarea about one-third as long as wide, slightly less than two-thirds as wide as the valve, divided by a moderately deep, narrow, subtriangular pedicle groove about one-tenth as wide as the pseudointerarea. Propareas flattened, slightly raised about the valve floor, with widely divergent flexure lines running close to their outer margins (Pl. 1, fig. 5). Dorsal valve about 98% as long as wide, gently convex with a blind umbonal area and a short, crescent-shaped pseudointerarea mainly occupied by the median groove, and narrow propareas slightly raised above the valve floor and bisected by flexure lines (Pl. 1, fig. 6). Shell ornamented by closely spaced, ridge-like concentric rugellae.

The details of the ventral and dorsal interiors are weakly impressed, except for the paired ventral umbonal muscle scars divided by indistinct pedicle nerve impression sometimes seen in the largest specimens. Characters of the muscle scars and mantle canal impressions are unknown.

Subfamily Glossellinae Cooper, 1956

Genus *Ectenoglossa* Sinclair, 1945

Type species (by original designation). – *Lingula lesueueri* Rouault, 1850, from the Grès Armoricain (Floian) of Brittany, France.

Ectenoglossa sorbulakensis Popov, 1980b

Plate 2, figure 8

1980b *Ectenoglossa sorbulakensis* Popov, p. 142, pl. 1, figs 1–4.

1984 *Ectenoglossa sorbulakensis* Popov; Nikitin & Popov *in* Klenina, Nikitin & Popov, p. 142, pl. 1, figs 1–4.

2002 *Ectenoglossa sorbulakensis* sp. nov., Popov; Popov, Cocks & Nikitin, p. 28, pl. 1, figs 1–4.

Holotype. – CNIGR 1/11523, ventral internal mould, from the Anderken Formation, Locality 1024, east side of Karatal valley, West Balkhash Region.

Material. – Berkytsyur Formation, Locality 817a: eight dorsal and ventral valves. Locality 1021: one ventral valve, NMW 98.28G.2233. Locality 8234: one ventral valve.

Remarks. – This species is probably the most widespread lingulide in the Sandbian of the Chu-Ili Terrane, where it occurs in numerous localities in the Anderken Formation (Popov *et al.* 2002) and is also known from the Bestamak Formation (upper Darriwilian – lower Sandbian) of the Chingiz Terrane (Nikitin & Popov *in* Klenina *et al.* 1984). It probably represents the major constituent of coquinas in the lingulide shell beds in sandstones of Unit 1 at the base of Baigara Formation; however, the preservation of these shells is not good enough for precise taxonomical identification and illustration.

Genus *Meristopacha* Sutton *in* Sutton *et al.*, 1999

Type species (by original designation). – *Lingula granulata* Phillips *in* Phillips & Salter, 1848, from the Golden Grove Group, Llandeilo Series (Sandbian), Carmarthenshire, Wales.

Remarks. – According to Sutton *et al.* (1999) *Pseudolingula spatula* Williams, 1974, selected by Popov & Cocks (2014) as the type species for the glossellinine genus *Holmeroglossa*, is a junior synonym of *Lingula granulata* Phillips *in* Phillips & Salter 1848. The latter species was selected as the type for *Meristopacha* Sutton *in* Sutton *et al.* 1999. Here we accept that suggested synonymy of these two species and therefore *Holmeroglossa* Popov & Cocks, 2014 should be considered as a junior objective synonym of *Meristopacha*.

Meristopacha sp.

Plate 1, figure 1; Plate 2, figures 11–16)

Material. – Kopgurgan Formation, Locality 388a: six ventral and six dorsal valves, including ventral valves NMW 98.28G.2145 (Lv, 5.6; W, 4.4; Pl. 2, fig. 11), 2146 (Lv, 4.8; W, 3.9; Pl. 2, fig. 12), 2150 (Lv, 6.5; W, 4.7; Il, 0.7; Iw, 1.4; Pl. 2, fig. 16), 2153, 2155, 2156, 2157, and dorsal valves NMW 98.28G. 2147 (Pl. 2, fig. 13), 2248 (Pl. 2, fig. 15), 2149 dorsal valves, 2251 (Ld, 4.1; W, 3.3; Sl, 3.3; Pl. 2, fig. 14), 2152, 2154.1, 2. Locality 8122: ventral valve NMW 98.28G.2162 (Pl. 1, fig. 1, 16.5; W, 10.2).

Description. – Shell subequally biconvex, elongate subtriangular, about 130% as long as wide with maximum width anterior to mid-length. Lateral margins gently rounded, anteriorly divergent, anterior commissure rectimarginate. Ventral valve very gently convex, subacuminate with pointed umbo. Ventral pseudointerarea narrow, subtriangular about half as long as wide and about 30% as wide as maximum shell width. Pedicle groove narrow with subparallel lateral margins. Ventral propareas not raised above the valve floor, lacking flexure lines. Dorsal valve gently convex, with a blind umbo, lacking pseudointerarea. Ventral interior with weakly impressed visceral area extending to mid-valve and defined anteriorly by the indistinct rim. A pair of subparallel pedicle nerve impressions fading towards the mid-valve and dividing posteriorly weakly impressed, paired umbonal muscle scars; other muscle scars and mantle canals not impressed. Dorsal interior with a strong median ridge terminated at about four-fifths valve length from the umbo, a pair of anterior lateral muscle scars situated on slightly thickened pads near the anterior termination of the dorsal median ridge. Weakly impressed gently divergent central muscle tracks extended slightly anterior to mid-valve. Other muscle scars and mantle canals not impressed.

Remarks. – These smooth and rather featureless lingulides are quite common in black shales of the Kopkurgan Formation. They are assigned to *Meristopacha* because they have a strongly elongate shell outline, a hemiperipheral growth of the dorsal valve which lacks a pseudointerarea, a strong dorsal median ridge extending anteriorly well beyond mid-length, and a vestigial ventral interarea with flexure lines on the propareas. However, they differ from *Meristopacha granulata*, which is the type and only named species of the genus, in having significantly smaller shell sizes, an unimpressed ventral visceral area and in the absence of distinct concentric rugellae. The Kazakh shells probably represent a separate species, but a new name is not erected because of the inadequate preservation of the available specimens.

Subfamily Elliptoglossinae Popov & Holmer, 1994

Genus *Elliptoglossa* Cooper, 1956

Type species (by original designation). – *Leptobolus*? *ovalis* Bassler, 1919, from the Martinsburg Shale (Sandbian) of Pennsylvania, USA.

Elliptoglossa magna Popov, 1977

Plate 2, figures 9, 10

1977 *Elliptoglossa magna* sp. nov. Popov, p. 104, pl. 24, figs 12–14.

Holotype. – CNIGR 5/10847, dorsal valve, from the Bestamak Formation (Sandbian), east side of Chagan River near Konur-Aulie cave, Chingiz Range, Kazakhstan.

Material. – Kopkurgan Formation, Locality 8122: three partly disarticulated shells, 16 ventral and dorsal valves, including BC 62411 (Pl. 2, fig. 9). Locality 8236, one dorsal valve NMW 98.28G.2132 (Pl. 2, fig. 10).

Remarks. – These shells differ from the specimens of *Elliptoglossa sylvanica* Cooper, 1956, which occur in the Uzunbulak Formation (Darriwilian) of the Chu-Ili Terrane (Nikitina *et al.* 2006), mainly in the presence of an incipient pedicle grove in the ventral valve and the almost smooth interiors of both valves, which lack distinct internal morphological features in spite of their large shell size (up to 6 mm length). The species was previously known from the Bestamak Formation (Sandbian) of the Chingiz Terrane (Popov 1977).

Family Elkaniidae Walcott & Schuchert *in* Walcott, 1908

Genus *Broeggeria* Walcott, 1902

Type species (by original designation). – *Obolella salteri* Holl, 1865, from the White-leaved Oak Shale Formation (Furongian); Malvern Hills, Herefordshire, England.

Broeggeria cf. *B. putilla* (Tenjakova, 1989)

Plate 1, figures 13, 14

2006 *Broeggeria* cf. *putilla* (Tenjakova, 1989); Nikitina, Popov, Neuman, Bassett & Holmer, p. 160, figs 14.19–14.22.

Material. – Tastau Formation, Locality 162: one ventral valve, NMW 98.28G.2129 (Pl. 1, fig. 13), and one dorsal valve NMW 98.28G.2131; Locality 163: one dorsal valve, NMW 98.28G.2130 (Pl. 1, fig. 14).

Remarks. – These shells are very similar to the specimens of *Broeggeria* cf. *B. putilla* described and illustrated by Nikitina *et al.* (2006) from the lower Uzunbulak Formation of the Kopalysai Sector in the southern Chu-Ili Range and may well be conspecific.

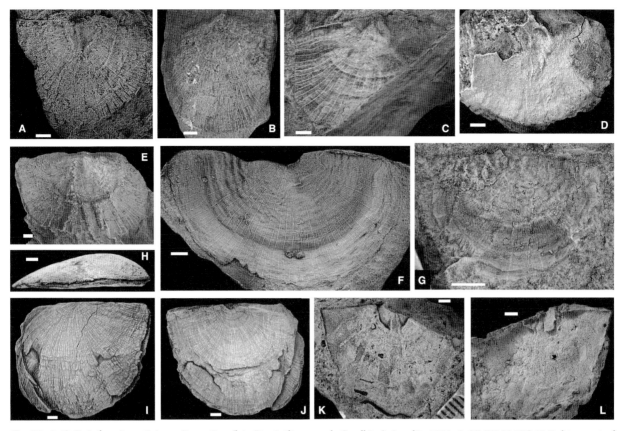

Fig. 17. **A–C, E.** *Infurca*? sp.; Baigara Formation (late Darriwilian – early Sandbian), Locality 1026; A, NMW 98.28G.1043, latex cast of dorsal interior; B, NMW 98.28G.1044, latex cast of ventral exterior; C, BC 60851b, ventral internal mould; E, NMW 98.28G.1045, dorsal external mould. **D,** *Colaptomena insolita* (Nikitina, 1985); Baigara Formation (late Darriwilian – early Sandbian), Locality 1021; NMW 98.28G.1379, dorsal interior. **F,** *Leptaena* (*Ygdrasilomena*) sp. Berkutsyur Formation (Sandbian), Locality 813; BC 62355, latex cast of ventral internal mould. **G,** *Buminomena* sp.; Kipchak Limestone (late Darriwilian – early Sandbian), Locality 166, NMW 98.28G.2034; dorsal exterior. **H–J.** *Katastrophomena rukavishnikovae* (Nikitina, 1985); Baigara Formation (late Darriwilian – early Sandbian), Locality 625/3; NMW 98.28G.897, conjoined valves, lateral, dorsal and ventral views. **K, L.** *Katastrophomena* sp., Kopkurgan Formation (Katian), Locality 735; BC 63876c, ventral internal mould and latex cast. Scale bars 2 mm.

Family uncertain

Genus *Oxlosia* Ulrich & Cooper, 1936

Type species (by original designation). – Eunoa accola Clarke, 1902, from the Levis Shale Formation (Middle Ordovician), Quebec, Canada.

Oxlosia? sp.

Plate 1, figure 12

Remarks. – This taxon is represented by a single, incomplete, elongate, suboval, almost flat dorsal valve, NMW 98.28G.2128 from Locality 154. Its apex is located at a short distance anterior to the posterior margin, ornamented by coarse concentric rugellae. In spite of its simple morphology, it is rather different from most Ordovician lingulides and can be compared only with the enigmatic *Oxlosia*, which is also characterised by an elongate shell with submarginal apex of both valves and coarse concentric surface ornament. However, the type species of *Oxlosia* remains inadequately known, and in the absence of data on the interiors of both valves, precise taxonomical attribution of the single shell from the Tastau Formation is not possible.

Superfamily Discinoidea Gray, 1840

Family Trematidae Schuchert, 1893

Genus *Trematis* Sharpe, 1848

Type species (by subsequent designation of Davidson (1853, p. 130). – Orbicula terminalis Emmons,

1842, from the 'Trenton Limestone' (Sandbian) of New York State, USA.

Trematis aff. *T. parva* Cooper, 1956

Plate 2, figure 6

1980   *Trematis* aff. *T. parva* Cooper, 1956; Nazarov & Popov, p. 113, pl. 31, figs 1, 2.

Material. – Baigara Formation, Locality 1022: CNIGR 215/11352, articulated shell. Locality 1026b, CNIGR 215/11352, ventral valve.

Remarks. – These specimens were briefly described and illustrated by Nazarov & Popov (1980). A ventral valve has the umbo situated at about one-third valve length from the posterior margin and a broad triangular pedicle notch covered by a small listrium posterior to the umbo. The dorsal valve is incompletely preserved, and the interiors of both valves are unknown. *Trematis parva* was described by Cooper (1956) from the Chatham Hill Formation (Sandbian) of Virginia, USA.

Trematis? sp.

Plate 2, figures 5, 7

Material. – Berkutsyur Formation, Locality 8234: one dorsal valve, BC 63865a.

Remarks. – The single specimen is characterised by a subcircular, gently convex dorsal valve with a marginal umbo and short curved posterior margin. The distinctive surface ornament of subhexagonal pits arranged in radial rows and gradually increasing in size suggests its attribution to *Trematis*; however, without data on the ventral valve morphology, this generic attribution cannot be firmly verified. This specimen is not conspecific with *Trematis?* sp. from the Anderken Formation (Sandbian) of Kopalysai in the Chu-Ili Range, briefly described and illustrated by Popov *et al.* (2002), because the latter has faint subcircular pits arranged in rows between distinct radial capillae. *Trematis* aff. *T. parva* Cooper, 1956, from the lower part of the Baigara Formation (late Darriwilian to early Sandbian) also has an ornament of polygonal pits, but they are smaller and are not arranged in radial rows.

Order Siphonotretida Kuhn, 1949
Superfamily Siphonotretoidea Kutorga, 1848
Family Siphonotretidae Kutorga, 1848

Genus *Multispinula* Rowell, 1962

Type species (by original designation). – *Schizambon macrothyris* Cooper, 1956, from the Wardell Limestone (Sandbian) of Virginia, USA.

Multispinula sp.

(Not illustrated)

1980   *Multispinula* sp. Nazarov & Popov, p. 118, pl. 30, fig. 5.

Material. – Baigara Formation, Locality 1022: CNIGR 250/11352, ventral valve.

Remarks. – This unnamed species is known from the single ventral valve described and illustrated by Nazarov & Popov (1980), and is the only documented occurrence of the genus in the Ordovician of the Kazakh terranes.

Class Craniata Williams, Brunton, Carlson, Holmer & Popov, 1996
Order Craniida Waagen, 1885
Superfamily Cranioidea Menke, 1828
Family Craniidae Menke, 1828

Genus *Petrocrania* Raymond, 1911

Type species (by original designation). – *Craniella meduanensis* Oehlert, 1888, from the Lower Devonian of Brittany, France.

Petrocrania? sp.

Plate 2, figure 1

Material. – Kopgurgan Formation, Locality 735: Two dorsal internal and external moulds, BC 63822b (Pl. 2, fig. 1), and BC 63880.

Remarks. – The specimens are the first and only craniides yet recorded from the Katian of all the Kazakh terranes. They are characterised by a low subconical dorsal valve with the umbo located slightly anterior of the posterior margin. The valve surface is ornamented by faint concentric fila and four to five thin concentric lamellae. There are no muscle scars and mantle canal impressions discernable on the internal moulds, which makes generic attribution of the shells questionable.

Genus *Pseudocrania* M'Coy, 1851

Type species (by subsequent designation of Davidson (1853) explanation of pl. 91). – *Orbicula antiquissima* Eichwald, 1840 [a junior synonym of *Crania petropolitana* Pander, 1830], from the Volkhov Formation (Dapingian to early Darriwilian) near St Petersburg, Russia.

Pseudocrania karatalensis Popov *in* Nazarov & Popov, 1980

(Not figured)

1980 *Pseudocrania karatalensis* sp. nov. Popov; Nazarov & Popov, p. 118, pl. 30, figs 6–8, text-fig. 60.

2013 *Pseudocrania karatalensis* Popov; Bassett, Ghobadi Pour, Popov & Kebria-ee Zadeh, p. 214, figs 3C, 4I–L.

Holotype. – CNIGR 251/11352, ventral valve; Baigara Formation (uppermost Darriwilian to lower Sandbian), Locality 1026b, west side of Karatal river, West Balkhash Region.

Material. – Locality 1026: five dorsal internal moulds, including NMW 98.28G.1988, CNIGR 251/11352, and CNIGR 251/11353.

Remarks. – This is the only known occurrence of *Pseudocrania* outside the Baltica palaeocontinent. The species was described in detail by Popov (*in* Nazarov & Popov 1980) and re-illustrated with brief notes by Bassett *et al.* (2013). While the generic affiliation of the Kazakh shells looks unquestionable, their age is considerably younger than that of the Baltoscandian species, which are confined to the Dapingian and early Darriwilian.

Class Chileata Williams, Brunton, Carlson, Holmer & Popov, 1996
Order Dictyonellida Cooper, 1956
Superfamily Eichwaldioidea Schuchert, 1893
Family Eichwaldiidae Schuchert, 1893

Genus *Eichwaldia* Billings, 1858

Type species (by original designation). – *Eichwaldia subtrigonalis* Billings, 1858, from the Ottawa Formation (Darriwilian) of Ontario, Canada.

Eichwaldia sp.

Plate 2, figures 2–4

Material. – Berkutsyur Formation (early Sandbian), NMW 98.28G.2110, ventral valve, from Locality 8121, area about 7 km south-west of Alakol Lake.

Remarks. – The single specimen has a subtriangular, evenly convex, slightly transverse ventral valve about 4 mm wide. The umbonal region is occupied by a large subtriangular opening covered by the colleplax and having a narrow slit along the anterior margin. It is similar to *Eichwaldia subtrigonalis*, which is the type and only known species of the genus yet described. However, *Eichwaldia* sp. differs from the type species in having finely pustulose shell ornament, previously unknown in dictyonellides. *Eichwaldia subtrigonalis* was described as having a smooth shell, but all the known Canadian shells of *Eichwaldia* are coarsely silicified and it is possible that any original micro-ornament has not been preserved in them.

Class Strophomenata Williams, Brunton, Carlson, Holmer & Popov, 1996
Order Strophomenida Öpik, 1934
Superfamily Strophomenoidea King, 1846
Family Strophomenidae King, 1846
Subfamily Strophomeninae King, 1846

Genus *Esilia* Nikitin & Popov, 1985

Type species (by original designation). – *Esilia tchetverikovae* Nikitin & Popov, 1985, from the

Andryushinskaya Formation (Sandbian), Ishim River basin, north-central Kazakhstan, Kalmykkol-Kokchetav Terrane.

Esilia cf. *E. tchetverikovae* Nikitin & Popov, 1985

Plate 2, figure 17; Plate 3, figures 1–5

cf. 1985 *Esilia tchetverikovae* sp. nov. Nikitin & Popov, p. 39, pl. 3, figs 1–9.

Material. – Takyrsu Formation, Locality 166: ventral valve NMW 98.28G.2111 (Pl. 2, fig. 17). Kopkurgan Formation, Locality 816: dorsal valve, BC 60825. Berkutsyur Formation, Locality 8121: ventral valve, BC 60783. Locality 8233: articulated shell, BC 62425 (Pl. 3, figs 1, 2); dorsal valve, BC 62426 (Pl. 3, figs 3–5). Locality 8236: ventral valve, BC 60782. Total one articulated shell, three ventral and two dorsal valves.

Remarks. – BC 60825 has a dorsibiconvex shell, a strongly uniplicate anterior commissure, relatively coarse multicostellate radial ornament with 5–6 ribs per 3 mm along the anterior margin, and a weak umbonal dorsal sulcus and a high dorsal median fold with steep lateral slopes originating slightly anterior from the beak. However, the conjoined valves BC 62425 from a different locality lack any fold or sulcus. The specimens are probably conspecific with *Esilia tchetverikovae* Nikitin & Popov, which was originally described from the Kupriyanovka Formation (early Sandbian) of the Ishim Region in northern Kazakhstan; however, the rather poorly preserved BC 62426 is the only interior known from our material and thus the species assignment here is provisional. *Esilia tchetverikovae* differs from *E. variabilis* (Nikitin & Popov, 1985), which occurs in the lower part of the Bestamak Formation (early Sandbian) of the Chingiz Range, Chingiz-Tarbagatai Terrane, in having a coarser radial ornament and in the absence of plications on the lateral sides of both valves.

Genus *Infurca* Percival, 1979

Type species (by original designation). – *Infurca tessellata* Percival, 1979, from the Goonumbla Volcanics (Katian), New South Wales, Australia.

Infurca? sp.

Figure 17A–C, E

Material. – Baigara Formation, Locality 1026: three ventral and one dorsal valves, including, NMW 98.28G.1043 (Fig. 17A); ventral valves, NMW 98.28G.1044 (Fig. 17B), 1045 (Fig. 17E), and BC 60851b (Fig. 17C).

Remarks. – The Balkhash specimens are poorly preserved, and the dorsal interior is known from a single composite external and internal mould; yet the dorsibiconvex shell, the open anteriorly ventral muscle field bounded laterally by short ridges, the character of the cardinalia and the absence of side septa in the dorsal valve all suggest affiliation to *Infurca*; however, they lack the crenulated surface of the teeth reported in *Infurca tessellata* Percival, 1979, which is the type and only yet named species of the genus. The latter taxon is also characterised by coarser, multicostellate radial ornament, unlike the parvicostellate Kazakh shells, and thus the generic identification is queried here.

Subfamily Furcitellinae Williams, 1965

Genus *Katastrophomena* Cocks, 1968

Type species (by original designation). – *Strophomena antiquata woodlandensis* Reed, 1917, from the Woodland Formation (Rhuddanian) of Girvan, Scotland.

Remarks. – *Kastrophomena* was widespread in the Late Ordovician and Silurian (Cocks & Rong 2000), but this species is the earliest known representative of that genus.

Katastrophomena rukavishnikovae (**Nikitina, 1985***)*

Plate 4, figures 2–9; Figure 17H–J

1985 *Strophomena rukavishnikovae* sp. nov. Nikitina, p. 27, pl. 2, figs 6–12.

Holotype. – CNIGR 32/12093, ventral internal mould from the Rgaity Formation (Darriwilian) of

Talapty Farm, southern Kendyktas Range, Kazakhstan.

Material. – Baigara Formation, Locality 1020: five articulated shells, including NMW 98.28G.846–847. Locality 1021: 25 articulated shells, including NMW 98.28G.865–890 and one ventral valve, NMW 98.28G.864 (Pl. 4, figs 6, 7). Locality 1022: 23 articulated shells, including NMW 98.28G.850–860 (Pl. 4, figs 3, 4), 861 (Pl. 4, fig. 2), 862 (Pl. 4, figs 8, 9), 863 (Pl. 4, fig. 5); Locality 1023: 8 articulated shells, NMW 98.28G.848–849, NMW 98.28G.891–896. Locality 625/3: articulated shell, NMW 98.28G.897 (Fig. 17H–J). Total 61 articulated shells and one ventral valve. Average sizes for articulated shells from localities 1022 and 1023. Length: X, 17.9; S, 3.39; min, 12.6; max, 28.9; N, 20. Width: X, 26.0, S, 3.56; min, 22.3; max, 32.2; N, 8. Thickness: X, 5.1; S, 1.36; min, 2.8; max, 7.4.

Description. – Shell resupinate, transverse, semioval to subrectangular, on average 71.9% (S, 6.7; OR, 65.3–89.8%; N, 7) as long as wide with maximum width at the hinge line or slightly anterior to it and on average 29.1% (S, 8.2; OR, 15.5–43.7%; N, 20) as thick as long. Cardinal extremities rectangular to slightly obtuse; anterior commissure rectimarginate. Ventral valve with moderately high (up to 3 mm) planar apsacline interarea. Delthyrium partly covered by a narrow convex pseudodeltidium. Dorsal valve lateral profile moderately and evenly convex with flattened umbonal area. Dorsal area low planar and anacline, with small, convex chilidium. Radial ornament multicostellate usually with 10–12 ribs per 3 mm (numbering 9–13 in 1, 8, 9, 4, 3 specimens). Concentric ornament of densely packed, elevated fila (6–10 per mm). Ventral interior with teeth supported by short widely divergent dental plates which continue as strong convergent muscle-bounding ridges which do not merge anteriorly in front of the ventral adductor muscle scars. Ventral muscle field large and subpentagonal. Ventral adductor muscle scars linear, bounded laterally by two almost parallel ridges. Ventral diductor scars large gently impressed, longer, but not enclosing adductor scars. Dorsal interior with strong, 'Type-A' double cardinal process (after Rong & Cocks 1994) with discrete, robust lobes situated upon a hinge line and occupying most of the low and narrow notothyrial platform (Nikitina 1985, pl. 2, figs 10b, 12). Socket plates short, blade-like, with their outer parts strongly curved laterally. Dorsal transmuscle ridges and median septum absent.

Remarks. – *Katastrophomena rukovishnikovae* was not listed among the earliest Middle Ordovician strophomenoids in the review of Zhan *et al.* (2013); however, it is of late Darriwilian age and, together with the *Strophomena*? sp. described by Laurie (1991) from the Cashions Creek Limestone (*Lepidomena fortimuscula* Zone) of Tasmania, is among the earliest representatives of the genus. The Australian species is very similar to *Katastrophomena rukovishnikovae* in its ventral muscle field and dorsal cardinalia.

Katastrophomena sp.

Figure 17K, L

Remarks. – A single ventral internal mould, BC 63876c from the Kopkurgan Formations (Katian) at Locality 735, shows a resupinate lateral profile, short, widely divergent dental plates merged with strong, curved, medially interrupted, muscle bounding ridges outlining a small subcircular muscle field ventral muscle field and finely impressed lemniscate mantle canals, all characteristic of *Katastrophomena*. However, in the absence of data on the dorsal interior and exterior of both valves, the species remains in open nomenclature.

Furcitellinae gen. et sp. indet.

Plate 3, figures 6–8

Material. – Baigara Formation, Locality 1022: three pairs of exfoliated conjoined valves, NMW 98.28G.1039 (Pl. 3, fig. 6), 1040 (Pl. 3, fig. 7), 1041 (Pl. 3, fig. 8).

Remarks. – This unidentified species has a relatively large, gently concavoconvex shell ornamented by strong, continuous, slightly undulating concentric rugellae. The radial ornament is finely parvicostellate, with 5–7 accentuated ribs originating at the umbo. The characteristic concentric ornament is similar to *Bellimurina* Cooper, 1956, but the species lacks geniculation and, since the interiors of both valves are unknown, generic determination is impossible and there is not enough material to erect a new species.

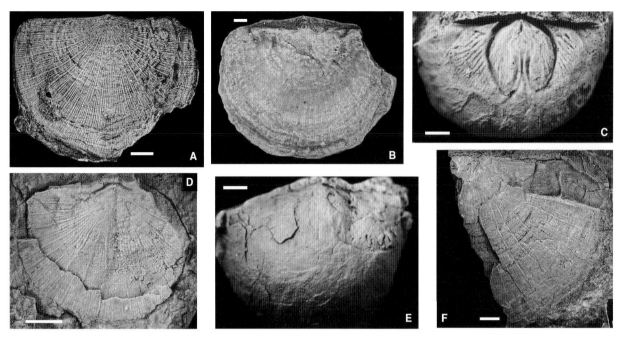

Fig. 18. **A–E.** *Buminomena* sp.; Berkutsyur Formation (Sandbian); **A,** Locality 812, BC 62331, latex cast of dorsal exterior; **B,** Locality 815, BC 62362, dorsal view of conjoined valves; **C,** Locality 812, BC 60843, ventral internal mould; **D,** Locality 8121, NMW 98G.2056, partly exfoliated dorsal exterior; **E,** Locality 815, BC 62361, partly exfoliated ventral exterior. **F,** Glyptomenidae indet.; Kopkurgan Formation (Katian), 8 km SW of Lake Alakol, Locality 727; BC 63815, partly exfoliated dorsal valve. Scale bars 2 mm.

Family Rafinesquinidae Schuchert, 1893
Subfamily Rafinesquininae Schuchert, 1893

Genus *Rhipidomena* Cooper, 1956

Type species (by original designation). – Strophomena tennesseensis Willard, 1928, from the Benbolt Formation (Darriwilian), Virginia, USA.

Rhipidomena sp.

Plate 4, figure 1

Remarks. – A single ventral internal mould, BC 63820f from the Kopkurgan Formations (Katian) at Locality 735, has a semicircular outline with maximum width at half length, an anteriorly resupinate sagittal profile, a delthyrium covered apically by a small pseudodeltidium, widely divergent dental plates, a large flabellate ventral muscle field without muscle bounding ridges, which are open anteriorly, all of which are characteristic of *Rhipidomena* (Cocks & Rong 2000). The specimen may be conspecific with *Rhipidomena* sp. from the Degeres Beds (mid Katian) of the Dulankara Mountains in the Chu-Ili

Range (Popov & Cocks 2006), but the latter is known only from one ventral valve and one dorsal internal mould.

Genus *Testaprica* Percival, 2009

Type species (by original designation). – Testaprica rhodesi Percival, 2009, from the Gunningbland Formation (Katian) of New South Wales, Australia.

Testaprica alperovichi n. sp.

Plate 3, figures 9–16

Derivation of name. – After the late E. V. Alperovich, Kazakh geologist, in appreciation of his work on the Palaeozoic of the West Balkhash Region.

Holotype. – BC 62344 (L, 16.3, W, 24.4; Pl. 3, figs 14, 15), a ventral internal mould from Locality 813, Berkutsyur Formation, area 4 km south-west of Lake Alakol, West Balkhash Region.

Material. – Locality 812: one pair of conjoined valves, two ventral and three dorsal valves, including BC 60845–49. Locality 813: two conjoined valves, 14

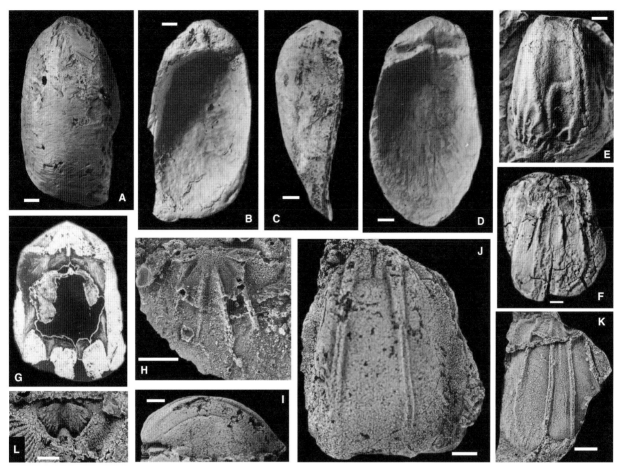

Fig. 19. **A–G, K.** *Christiania* cf. *C. tortuosa* Popov, 1980a; **A–G,** Baigara Formation (late Darriwilian – early Sandbian); **K,** Berkutsyur Formation (Sandbian); **A–C,** Locality 1026b, NMW 98.28G.1029, conjoined valves, ventral, dorsal and side views; **D,** Locality 1026b, NMW 98.28G.1030, conjoined valves, ventral view; **E,** Locality 1026, NMW 98.28G.1042, ventral internal mould; **F,** Locality 1026b, NMW 98.28G.1031, dorsal interior; **G,** Locality 1026b, NMW 96.28G.1032, acetate replica from section through an articulated shell parallel to the commissural plane showing cardinal process, median and two pairs of side septa; **K,** Locality 815, BC 62367, latex cast of dorsal interior. **H–J,** *Christiania* cf. *C. proclivis* Popov & Cocks, 2006; Kopkurgan Formation (Katian), Locality 735; **H,** BC 63822a, latex cast of incomplete dorsal interior; **I, J,** BC 63884, ventral internal mould, side and ventral views. **A–K.** Scale bars 2 mm. **L.** *Leangella* (*Leangella*) sp., Kopkurgan Formation (Katian), Locality 735; BC 63824d, dorsal internal mould, scale bar 1 mm.

ventral and 9 dorsal valves, including, BC 62342 (Pl. 3, fig. 10), BC 62343 (L, 22.0 mm; Pl. 3, figs 11, 16), BC 60726–31, BC 62341 (Pl. 3, figs 12, 13), NMW 98.28G.1228 (Pl. 3, fig. 9), 1229. Locality 814: two conjoined valves, eight ventral and three dorsal valves, including BC 60747–54 and BC 62358b.

Diagnosis. – *Testaprica* with a distinctly apsacline ventral interarea, finely parvicostellate radial ornament, muscle bounding ridges completely enclosing diamond-shaped ventral muscle field and rudimentary dorsal median septum.

Description. – Shell convexoconcave, transverse, semioval, about 70% as long as wide with maximum width at or slightly anterior to the hinge line. Cardinal extremities slightly acute to almost rectangular.

Anterior commissure rectimarginate, broadly rounded. Ventral valve evenly concave with a low, planar, apsacline interarea. Delthyrium small and triangular covered by convex pseudodeltidium. Dorsal valve profile evenly convex with maximum height at mid-length. Dorsal interarea low, apsacline with a small convex chilidium. Radial ornament parvicostellate with 1–4 parvicostellae between accentuated ribs and with 7–9 ribs per 2 mm along the anterior margin. Concentric ornament of fine, evenly spaced fila, about 8 per mm. Ventral interior with small, transverse teeth supported by thin, divergent dental plates continuing as high muscle-bounding ridges completely enclosing the ventral muscle field. Ventral muscle field strongly impressed, subpentagonal, about one-third as long as the valve. Adductor scars lenticular, bisected posteriorly by a faint median ridge. Diductor scars subtriangular, slightly longer but not enclosing the adductor scars. Mantle canals poorly impressed except *vascula genitalia* situated laterally from the ventral muscle field (Pl. 3, fig. 14). Dorsal interior with delicate, bilobed 'Type B' (after Rong & Cocks 1994) cardinal process with slightly elongate, entirely discrete lobes merged with the almost straight, transverse socket plates. A pair of prominent, subparallel side septa extend to the mid-valve. Rudimentary median septum present in gerontic individuals as a faint, short ridge in the anterior part of the adductor muscle field. Adductor muscle scars narrow, lenticular, divided by side septa. Dorsal mantle canals lemniscate with very short divergent *vascula media* bifurcating into two major branches immediately in front of the adductor muscle field. *Vascula myaria* transverse, bounded anteriorly by papillose impressions of *vascula genitalia* (Pl. 3, figs 11, 16).

Remarks. – *Testaprica alperovichi* differs from the type species *Testaprica rhodesi* in having parvicostellate (not multicostellate) radial ornament, a distinctly apsacline ventral interarea, and longer dorsal side septa, which extend anteriorly to the mid-valve. It differs from another Kazakh species *Testaprica ajaguzensis* (Borissiak, 1955), which occurs in the middle to late Katian Akdombak Formation of the Chingiz Range (Popov & Cocks 2014), in having finer radial ornament, a rudimentary dorsal median ridge, muscle bounding ridges completely enclosing the ventral muscle field, as well as in the absence of the ventral subperipheral rim and oblique rugellae along the dorsal posterior margin.

Genus *Colaptomena* Cooper, 1956

Type species (by original designation). – *Colaptomena leptostrophoidea* Cooper, 1956, from the Martinsburg Formation (Sandbian), Virginia, USA.

Colaptomena insolita (Nikitina, 1985)

Plate 3, figures 17–23; Figure 17D

1985 *Macrocoelia insolita* sp. nov. Nikitina, p. 26, pl. 2, figs 1–5.

Holotype. – CNIGR 32/12093, ventral internal mould from the Rgaity Formation (Late Darriwilian) of Talapty Farm, southern Kendyktas Range, Kazakhstan.

Material. – Baigara Formation, Locality 1020a, conjoined valves. Locality 1021: 12 conjoined valves and one dorsal valve, including NMW 98.28G.995–1028, 1378 (Pl. 3, figs 20, 21), 1379 (Text-fig. 17D), 1380, 1381, 2101-2108. Locality 1022: 57 conjoined valves, including NMW 98.28G.924 (Pl. 3, figs 22, 23), 925–994. Locality 1023: 12 conjoined valves, including NMW 98.28G. 2058-2059. Locality 625/3: four conjoined valves, including BC 60864–67. Rgaity Formation (North Tien Shan), Locality 11221: five ventral valves and seven dorsal valves, including NMW 98.28G.2012 (Pl. 3, fig. 19), 2013 (Pl. 3, fig. 17), 2075–2085, 2114, 2115. Total 85 articulated shells, four ventral and two dorsal valves. Average sizes for articulated shells from Localities 1022. Length: X, 18.0; S, 3.24; min, 10.4; max, 22.6; N, 29. Width: X, 23.0, S, 3.51; min, 17.7; max, 30.2; N, 25. Thickness: X, 3.9; S, 0.85; min, 2.7; max, 6.5.

Description. – Shell planoconvex to weakly concavoconvex, transverse, semioval in outline, on average 78.2% (S, 14.0; OR, 46.0-98.4; N, 25) as long as wide, with maximum width at or immediately anterior to the hinge line and 22.2% (S, 6.6; OR, 15.3–46.2%; N, 25). Cardinal extremities almost rectangular to slightly acute. Anterior commissure rectimarginate. Ventral valve lateral profile gently convex with maximum height one-third from the umbo. Ventral interarea low, planar, apsacline with small apical pseudodeltidium. Dorsal valve almost flat to very gently concave with anacline dorsal interarea and a small chilidium. Radial ornament slightly unequally multicostellate with 8–11 ribs per 3 mm (numbering 6–12 in 2, 8, 18, 18, 9, 10, 4 specimens). Concentric ornament of fine, evenly spaced fila and 3–5 oblique rugellae along the hinge line. Ventral interior with small teeth and low, widely divergent dental plates. Muscle field weakly impressed, subtriangular, open anteriorly. Dorsal interior with a massive 'Type B' (of Rong & Cocks 1994) cardinal process with discrete, elongate, plate-like lobes situated anterior to the hinge-line and a short, straight, widely divergent socket plates (Nikitina, 1985, pl. 2, fig. 3b). Long faint median septum extending anteriorly to mid-valve. Adductor muscle scars weakly impressed crossed by two pairs of thin variably developed side septa. Low dorsal subperipheral rim present.

Remarks. – The specimens from the Baigara Formation are conspecific with the shells described by Nikitina (1985) as *Macrocoelia insolita* from the Rgaity Formation of the South Kendyktas Range, North Tien Shan Terrane. The minor differences, which include coarser and more uneven radial ornament, are within the range of intraspecific variation. In their revision of the group, Rong & Cocks (1994) established that the genus *Macrocoelia* is a junior synonym of *Colaptomena*.

Subfamily Leptaeninae Hall & Clarke, 1894

Genus *Leptaena* Dalman, 1828

Type species (by subsequent designation of King 1846, p. 28). – Leptaena rugosa Dalman. 1828, from the Dalmanitina Beds (Hirnantian) of Västergötland, Sweden.

Subgenus *Leptaena (Ygdrasilomena)* Cocks, 2005

Type species (by original designation). – Leptaena (Ygdrasilomena) roomusoksi Cocks, 2005, from the Boda Limestone (Katian) of Dalarna, Sweden.

Leptaena (Ygdrasilomena) sp.

Figure 17F

Material. – Berkutsur Formation, Locality 813: ventral valve, BC 62355 (Lv = 14.3, W = 33.2, Fig. 17F). Locality 8124: dorsal valve, BC 62419 (Ld = 6.5; West. = 10).

Remarks. – A ventral valve, BC 62355 from Locality 813 (Fig. 17F), shows ventral and dorsal geniculation, parvicostellate radial ornament, and a disc with concentric ornament of undulating rugae interrupted by accentuated ribs, all characteristic of *Leptaena (Ygdrasilomena)*. The left side of the specimen shows extensive but repaired damage from unsuccessful predation: the part of the shell anterior to the damage has lost its concentric ornament, but the characteristic undulating rugae are preserved on the right side. In addition, a partly exfoliated dorsal valve, BC 62419 from Locality 8124, is probably also the same taxon, although the ornament is poorly preserved. Two unnamed species previously referred to *Limbimurina* were reported from the Late Ordovician of Kazakhstan (Nikitin & Popov 1996; Popov *et al.* 2002), both from the Chu-Ili Terrane, but they have a strong interference pattern of oblique rugae on the

disc which are not interrupted by radial ornament and are therefore here re-assigned to *Leptaena (Ygdrasilomena)*.

Genus *Doughlatomena* n. gen.

Type species. – Doughlatomena splendens n. sp., see below.

Derivation of name. – After the Doughlat clan of Kazakhs.

Diagnosis. – Finely parvicostellate leptaenines with short teeth and weakly-impressed subquadrangular ventral muscle field; small erect cardinal process anterior to strong slightly curved socket plates flaring at more than right angles.

Discussion. – Doughlatomena is in many respects reminiscent of *Limbimurina* Cooper, 1956, but lacks the distinctive small rugellae of the latter genus, which also has more elated ventral muscle-bounding ridges which surround a suboval rather than subquadrangular muscle field. However, the dorsal valve interiors of the two genera are rather similar, with both possessing widely divergent and slightly curved socket plates. The only other Sandbian to Katian leptaenines, in addition to *Limbimurina*, are *Leptaena* itself and its other subgenera *Septomena* and *Ygdrasilomena* (see above), which all have more regular rugae and smaller socket plates; *Kiaeromena* Spjeldnæs, 1957, which has a much more gentle geniculation and much smaller socket plates than *Doughlatomena*; and the endemic Chinese *Hingganoleptaena* Zhu, 1985, which has distinctive incurved socket plates; all the other leptaenine genera are of Silurian or Devonian ages (Cocks & Rong 2000).

Doughlatomena splendens n. sp.

Plate 5, figures 1–11

Derivation of name. – From Latin *splendens,* magnificent.

Holotype. – BC 63873a (Pl. 5, figs 3, 4), dorsal valve internal mould, from the Kopkurgan Formation (Katian), Locality 735, about 10 km south-west of Lake Alakol, West Balkhash Region.

Diagnosis. – As for the genus.

Material and dimensions. – Locality 735: two artic-ulated shells, four ventral and 16 dorsal valves pre-served as decalcified shells, including BC 63821c (L, 22.0; W, 27.1), BC 63889b (L, 19.6; W, 19.8), BC 63819e, BC 63820e, BC 63823 (Pl. 5, figs 5, 6), BC 63824c (Pl. 5, fig. 11), BC 63832 (Pl. 5, fig. 9), BC 63885c; internal moulds, BC 63825c, BC 63849 (Pl. 5, fig. 2), (average L 18.8, range 7.6–23.1; average W 28.8, range 9.5–41.6, average L/W ratio 69.4, range 53.1–81.2): dorsal valves, external moulds, BC 63819f, BC 63824b (Pl. 5, fig. 10), BC 63825b, BC 63835c, BC 63840c, BC 63846c, BC 63888d (Pl. 5, fig. 1); internal moulds, BC 63819 g, BC 63834b, BC 63847c (Pl. 5, figs 7, 8), BC 63888e (average L 24.0, range 15.0–22.8; average W 32.1, range 24.4–33.0, average L/W ratio 74.8, range 54.1–89.8).

Description. – Semicircular outline, with alae very variably developed, sometimes elongate (maximum width at hinge line), sometimes absent (with maxi-mum width at 20% valve length). Shells initially gently biconvex to flat, but substantial trail devel-oped, commencing with variably raised ventral rim and progressing with a geniculation which is often not initially sharply developed but which ends up by varying from about 70° to right angles. Ornament of rather irregular and not very prominent rugae, which are lacking on the trail, as well as fine parvicostellae. Apsacline ventral interarea with the delthyrium cov-ered apically by the small pseudodeltidium; relatively small, anacline dorsal interarea with delthyrium cov-ered by an entire chilidium (seen on BC 63821). Ven-tral interior with widely divergent dental plates, a subquadrangular but bilobed anteriorly muscle field extending to about one-third valve length, bounded laterally by variably developed subparallel muscle bounding ridges and open anteriorly in some speci-mens, but with a slightly impressed flabellate anterior margin in others. Lemniscate mantle canal system not seen in most shells, but occasionally impressed in others (e.g. BC 63849: Pl. 5, fig. 2). Dorsal interior with erect plate-like cardinal process lobes of med-ium size which diverge from each other at about 40°, forming the posterior end of a broad but short notothyrial platform from which a pair of substantial curved socket plates originate, which initially diverge at about right angles but which curve round to end up at over one-third valve width and almost subpar-allel with the hinge line. The centre of the notothyr-ium extends for only approximately a fifth to a quarter of valve length before merging with the valve floor. A very short thin median septum is weakly developed just posterior to the valve centre, as well as a rather weak pair of side septa which diverge at about 30°. In some specimens, including the holo-type, the bilobed pear-shaped muscle field is weakly impressed on either side of the median septum. The anterior ends of the many small mantle canals are impressed on the anterolateral part of the trail only.

Remarks. – The most notable intraspecfic variable feature is the very different development of alae seen in the population, which varies from individuals with extensive alae and others with almost none.

Family Glyptomenidae Cooper, 1956

Remarks. – Glyptomenids described from Kaza-khstan are *Buminomena* (see below), two species of *Glyptomena*, and one described species, *Platymena tersa* Popov & Cocks and other probable records (e.g. *P.?* sp. of Popov & Cocks (2014) of *Platymena*, as well as the *Glyptomenoides?* sp. discussed by Popov & Cocks (2006).

Genus *Buminomena* Popov & Cocks, 2014

Type species (by original designation). – *Oepikina abayi* Klenina, *in* Klenina *et al.* 1984, revised by Popov & Cocks (2014), from the Taldyboi Formation (Katian) of the Chingiz-Tarbagatai Terrane, Kaza-khstan.

Buminomena sp.

Figures 17G, 18A–E

Material. – Berkutsyur Formation, Locality 812: one ventral internal mould, BC 60843 (Lv, 13.3; W, 15.7; T, 7.0; Ml, 7.2; W, 6.1; Fig. 18C); one dorsal external mould, BC 62331 (Fig. 18A). Locality 813: three dorsal valves. Locality 815: two articulated shells, including BC 62360, BC 62362 (Fig. 18B); one ventral valve, BC 62361 (Fig. 18E). Locality 816: dor-sal external mould, NMW 98.28G.2035. Locality 8121: two dorsal valves NMW 98.28G.2056 (Fig. 18 D); Locality 166: dorsal valve, NMW 98.28G.2034 (Fig. 17G). Total three articulated shells, two ventral and six dorsal valves.

Remarks. – The specimens are referred to *Bumi-nomena* since they have a strongly concavoconvex shell with the dorsal valve geniculate, finely parvicostellate radial ornament becoming multicostellate peripherally, and a bilobed muscle field surrounded by strong muscle bounding ridges except anteromedially. The dorsal interior is known from a single exfoliated specimen, which shows the characteristic for *Buminomena* long, straight, divergent socket plates, although the median and side septa are apparently lacking. Details of the cardinal process are unknown. The specimens show distinct similarity to *Buminomena abayi* (Klenina, *in* Klenina *et al.* 1984) in the proportions and size of the shell, radial ornament with up to 7 parvicostellae per mm, and a ventral valve interior with a bilobed muscle field surrounded by strong muscle bounding ridges; and having narrow adductor scars almost twice as short as diductor scars, and fine, divergent *vascula media*. However, the differences are a hinge line shorter than maximum shell width, and obtuse cardinal extremities in *Buminomena* sp., though the dorsal valve interior of the latter is inadequately known.

Glyptomenidae gen. et sp. indet.

Figure 18F

Remarks. – A single incomplete dorsal valve, BC 63815 from Locality 727 in the Berkutsyur Formation, is almost flat with finely parvicostellate radial ornament with up to 15 ribs per 3 mm along the anterior margin and superimposed fine, regular concentric fila. Among the Late Ordovician strophomenoids documented from the Chu-Ili Terrane, it appears most similar to *Glyptomena onerosa* Popov, 1980b, but definitive identification cannot be made on the sparse material.

Family Christaniidae Williams, 1953

Genus *Christiania* Hall & Clarke, 1892

Type species (by original designation). – *Leptaena subquadrata* Hall, 1883, from the Lenoir Formation (Darriwilian) of Tennessee, USA.

Remarks. – Many species of *Christiania* have been described from Kazakhstan, *C. hastata* (Rukavishnikova, 1956) revised by Nikitina *et al.* (2006) from the Uzunbulak Formation, *C. proclivis* Popov &

Cocks (2006; 2014) from the Dulankara and probably Akdombak Formations, *C. egregia* Popov (1980b) and Popov *et al.* (2002) from the Anderken Formation, *C. taldyboyensis* Klenina (*in* Klenina *et al.* 1984) from the Taldeboi Formation, as well as *C.* aff. *C. skolia* Percival, 1991, from the Betpak-Dala mudmound (Nikitin & Popov 1996) and *C.* aff. *sulcata* Williams, 1962 (Popov *et al.* 2002) from the Anderken Formation. That is in addition to the more than thirty species of the genus described from Middle to Late Ordovician rocks in many parts of the world.

Christiania cf. *C. proclivis* Popov & Cocks, 2006

Figure 19H–J

cf. 2006 *Christiania proclivis* sp. nov. Popov & Cocks, p. 262, pl. 2, figs 7–11.

Material. – Kopkurgan Formation, Locality 735: a pair of conjoined valves, four ventral and six dorsal valves, preserved as external and internal moulds, including BC 63819c, BC 63822a (Fig. 19H), BC 63830b, BC 63830d, BC 63837b, BC 63840d, BC 63845, BC 63873b, BC 63884 (Figs 19I, J).

Remarks. – These specimens are similar to *Christiania proclivis* from the Degeres Beds (middle Katian) of Dulankara Mountains, whose holotype is BC 57759, in the lateral profile of their strongly concavo-convex shells with steep lateral sides to the ventral valve, the absence of radial ornament, arrangement of muscle scars and mantle canals in the ventral valve, and strong side septa in the ventral valve, but there is no ventral sulcus. The dorsal valves available are incomplete, so the species attribution of the Baigara material is provisional. Discussion of the affinities of *Christiania proclivis* was given by Popov & Cocks (2006).

Christiania cf. *C. tortuosa* Popov, 1980b

Figure 19A–G, K

cf. 1980b *Christiania tortuosa* sp. nov. Popov, p. 56, pl. 17, figs 10–12.

Holotype. – CNIGR 1/11098, dorsal internal and external mould, from the Lidievka Formation (Sandbian) of Belyi Kardon, north-western Central Kazakhstan.

Material. – Baigara Formation, Locality 1026: five ventral and three dorsal internal moulds, including NMW 98.28G.1038, 1042 (Fig. 19E). Locality 1026b: 11 articulated shells and one dorsal valve, including NMW 98.28G.1029 (Fig. 19A–C), 1030 (Fig. 19D), 1031 (Fig. 19F), 1032 (Fig. 19G), 1033–1037. Berkutsyur Formation, Locality 812: three ventral and two dorsal internal moulds, one dorsal external mould, including BC 60844, NMW 98.28G.2054. Locality 815: dorsal interior, BC 62367 (Ld, 11.3; West, 9 mm; Fig. 19K). Total 11 articulated shells, eight ventral and eight dorsal valves. Average sizes for nine articulated shells from Locality 1026b. Length: X, 11.4; S, 2.34; min, 10.7; max, 17.7. Width: X, 9.0; S, 0.92; min, 7.0; max, 10.1. Thickness: X, 5.9; S, 1.40; min, 3.6; max, 9.3.

Description. – Shell concavoconvex, elongate, sub-rectangular, on average 150% (S, 21; OR, 121–182%; N, 9) as long as wide with maximum width at or slightly anterior to mid-length and 44% (S, 7; OR, 32–55%; N, 9) as thick as long. Cardinal extremities almost rectangular, slightly alate. Ventral valve strongly convex with maximum thickness at about one-third valve length and steep lateral slopes. Ventral umbonal area strongly swollen. Ventral inter-area strongly apsacline to almost orthocline with a large convex pseudodeltidium. Dorsal valve gently concave with a hypercline interarea and a complete, convex chilidium. Radial ornament finely multi-costellate, rarely preserved. Ventral interior with strong, transverse teeth and low, widely diverging dental plates. Muscle field weakly impressed and bilobate. Ventral mantle canals lemniscate with long, strongly impressed, almost parallel *vascula media*. Dorsal interior with bilobed cardinal process, low, curved socket plates, two pairs of strong side septa and a faint median septum, about half as long as the valve. Adductor scars bordered anteriorly by strong, transverse, posteriorly curved, muscle bounding ridges.

Remarks. – This taxon differs from *Christiania tortuosa* Popov, 1980b, which is widespread across the Chu-Ili Terrane in the late Sandbian Anderken Formation (Popov *et al.* 2002), in its shorter dorsal median ridge, which does not usually extend anteriorly to mid-length, rectimarginate anterior commissure, and in the absence of a ventral sulcus. Shells from the upper part of the Ichkebash Formation (Sandbian) of North Kyrgyzstan, described by Misius (1986) as *Christiania tenuicincta* (M'Coy, 1846), have a shallow ventral sulcus and are probably *Christiania tortuosa*.

Superfamily Plectambonitoidea Jones, 1928

Family Plectambonitidae, Jones, 1928

Subfamily Taphrodontinae Cooper, 1956

Genus *Bandaleta* Nikitin & Popov, 1996

Type species (by original designation). – *Bandaleta plana* Nikitin & Popov, 1996, from the Dulankara mud mound (Darriwilian), Betpak-Dala desert, Chu-Ili Terrane, Kazakhstan.

Bandaleta sp.

Plate 6, figures 1, 2; Plate 7, figure 14

Material. – Kopkurgan Formation, Locality 816: one articulated shell, six ventral and three dorsal valves, including NMW 98.28G.2057 (Pl. 6, fig. 2), 2165 (Pl. 7, fig. 14). Berkutsyur Formation, Locality 8120: two ventral valves, NMW 98.28G.1873 and 1874. Locality 8233: three articulated shells two ventral and three dorsal valves, including BC 62427–33. Locality 8124: three dorsal valves, BC 62420, BC 62421 (Ld, 12.0; West, 20 mm), BC 62422, (Pl. 6, fig. 1, Ld, 10.6, W, 22.3), BC 62432, Locality 8236a: ventral valve, BC 62438. Total four articulated shells, 11 ventral and eight dorsal valves.

Remarks. – A few specimens from the lower part of the Berkutsyur Formation are identified as *Bandaleta* because they have a planoconvex shell with parvicostellate ornament, whereas a dorsal valve reveals cardinalia with a simple blade-like cardinal process, short widely divergent socket plates, a double septum and a subquadrate adductor muscle field completely enclosed by the muscle bounding ridges typical of the genus (Pl. 6, fig. 1, Pl. 7, fig. 14). However, the identification remains uncertain because of inadequate preservation of the specimens. *Bandaleta plana* Nikitin & Popov, 1996, from the mud mound in the Betpak-Dala Desert, is the only formally named species of the genus in the Chu-Ili Terrane, but *Bandaleta* cf. *B. plana* was identified from the Dulankara Formation by Popov & Cocks (2006). *Taphrodonta bicornigera* described by Nikitin (1974) from the Sarybidaik Formation (Sandbian) of Narulgen, in the Boshchekskul Terrane of Kazakhstan was subsequently reassigned by Nikitin & Popov (1996) to *Bandaleta*. Both species are similar to the shells from the Berkutsyur Formation in size, proportions, and outline of the shell, characters of cardinalia with a simple cardinal process and short, widely divergent brachiophores, and a dorsal adductor field bounded anteriorly and laterally by distinct muscle bounding ridges, yet none of them is probably conspecific. However, shells in our collection differ from *Bandaleta plana* in having a weakly defined sulcus fading anterior to mid-valve, a less robust double septum and a weakly defined dorsal peripheral rim. They also have strongly differentiated parvicostellate radial ornament, unlike *Bandaleta bicornigera*.

Genus *Isophragma* Cooper, 1956

Type species (by original designation). – *Isophragma ricevillense* Cooper, 1956, from the Athens Formation (Sandbian), Tennessee, USA.

Isophragma princeps Popov, 1980b

Plate 6, figures 3–8

1980b *Isophragma princeps* sp. nov. Popov, p. 54, pl. 7, figs 1–5.

Holotype. – CNIGR 1/11098, dorsal internal and external mould, from the Lidievka Formation (Sandbian) of the Stepnyak Terrane at Belyi Kardon, north-western Central Kazakhstan.

Material. – Berkutsyur Formation, Locality 812: one articulated shell, three ventral and three dorsal valves, including BC 60811, BC 62329-30, BC 62332 (Pl. 6, fig. 3), BC 62333 (Pl. 6, fig. 5), BC 62350 (Lv, 6.7; West, 14; Pl. 6, figs 6, 7), BC 63699. Locality 813: articulated, three ventral and three dorsal valves, including BC 60823, BC 60824 (Lv, 8.8; W, 12.2; Pl. 6, fig. 8), BC 62349a (Pl. 6, fig. 4), BC 62351. Locality 815: dorsal external mould, BC 62366.

Description. – Shell resupinate, transverse, semioval, about three-quarters as long as wide with maximum width at hinge line. Cardinal extremities slightly acute to almost rectangular. Anterior commissure rectimarginate. Ventral valve lateral profile gently concave in the anterior half and slightly convex posterior to mid-length with maximum height at the umbonal area. Ventral interarea planar, apsacline with a narrow, convex pseudodeltidium. Dorsal lateral profile gently concave to almost flat posteriorly, becoming convex anterior to the mid-length, with maximum height one-third valve length. Dorsal interarea low,

cataclime to slightly hypercline with a narrow, convex chilidium. Shallow dorsal sulcus originating at the umbonal area and fading near the mid-valve. Radial ornament multicostellate with 7–8 ribs per 2 mm at the anterior margin. Concentric ornament of faint, densely spaced growth lines. Ventral interior with strong teeth and no dental plates. Ventral muscle field about two-fifths valve length, with subtriangular, strongly raised anteriorly adductor muscle scar and strongly impressed, elongate and divergent diductor scars about one-third longer than the adductor scar. Subperipheral rim high, with median gap in front of the adductor scar. Ventral mantle canals saccate with straight divergent *vascular media* and *vascula arcuata* following the margin of the subperipheral rim. Dorsal interior with a simple bulbous cardinal process, short, straight, widely divergent socket plates and deep, transverse sockets. Double septum high, subtriangular, originating in front of the cardinal process, about four-fifths valve length. Adductor scars subquadrate, extending anteriorly to the mid-valve, surrounded by high muscle bounding ridges, merged anteriorly with a high subperipheral rim.

Remarks. – This species differs from *Isophragma imperator* Popov, 1980b, which is the only other known Kazakh species of the genus, from the late Sandbian Anderken Formation of the Chu-Ili Terrane, in having a rectimarginate anterior commissure, double septa not merged at the base, and a strong dorsal subperipheral rim.

Family Bimuriidae Cooper, 1956

Genus *Bimuria* Ulrich & Cooper, 1942

Type species (by original designation). – *Bimuria superba* Ulrich & Cooper, 1942, from the Arline Formation (Darriwilian) of Tennessee, USA.

Bimuria karatalensis n. sp.

Plate 6, figures 9–19

Derivation of name. – After the river Karatal near the type locality.

Holotype. – NMW98.28G.1122 (Pl. 6, fig. 10), ventral internal mould, from the Baigara Formation (late Darriwilian–early Sandbian), from Locality 1026, west side of Karatal River, southern Betpak-Dala; Kazakhstan.

Material. – Baigara Formation: Locality 1026: seven ventral and seven dorsal valves preserved as external and internal moulds, including; NMW 98.28G.1046 (Pl. 6, fig. 11), 1047 (Pl. 6, fig. 9), 1123 (Pl. 6, fig. 19), 1124. Locality 1026b: 26 conjoined valves, 14 ventral and 6 dorsal valves, including NMW 98.28G.1046, 1048-21 (Pl. 6, fig. 15), 1125 (Pl. 6, figs 13, 14), 1126 (Pl. 6, figs 16, 17). Kopkurgan Formation: Locality 816: a pair of conjoined valves, BC 62379 (Pl. 6, figs 12, 16). Average sizes for articulated shells from Locality 1026b. Length: X, 18.8; S, 3.20; min, 15.7; max, 26.1; N, 16. Width: X, 26.5; S, 3.81; min, 16.6; max, 31.1, N, 14. Thickness: X, 8.3; S, 1.77; min, 6.5; max, 12.6; N, 16.

Diagnosis. – Shell large for the genus, transverse, semioval. Ventral valve sagittal profile evenly convex, Dorsal side septa faint; bema weakly defined laterally and anterolaterally by a rim, not undercut outwards.

Description. – Shell concavoconvex, transverse, semioval, on average 70.7% (S, 12.6; OR, 66.1–107.2%; N, 14) as long as wide with maximum width at a hinge line and 44.1% (S, 7.9; OR, 34.5–58.6%; N, 16) as thick as long. Cardinal extremities slightly acute to almost right angle; anterior commissure rectimarginate. Ventral valve strongly and evenly convex with a swollen umbonal area. Ventral interarea low, planar, anacline, with apical pseudodeltidium. Dorsal valve evenly concave with low, hypercline interarea. Notothyrium with discrete chilidial plates. Shell surface with comae which are more strongly developed in the ventral valve. Ventral interior with strong transverse teeth. No dental plates. Ventral muscle field weakly impressed, open anteriorly. Ventral mantle canals lemniscate with long almost parallel vascula media. Dorsal interior with a simple cardinal process and low, short, widely divergent socket plates. Bema divided, elongate, very weakly defined, bisected. Side septa thin, only slightly stronger than the median septum.

Remarks. – In external morphology, *Bimuria karatalensis* recalls the type species *Bimuria superba* as well as *Bimuria buttsi* Cooper, 1956, which both occur in the Little Oak Formation (Sandbian) of Alabama, but it can be distinguished from those two by its weakly developed bema and faint side septa in the dorsal valve. It differs from the only named species from Kazakhstan, *Bimuria triquetra* Nikitin & Popov (*in* Klenina *et al.* 1984) from the Bestamak Formation (Sandbian) of the Chingiz-Tarbagatai Terrane in having an almost twice as large, suboval (not subtriangular) shell, evenly convex sagittal profile of ventral valve, and a weakly defined rim bounding the bema laterally and anterolaterally, while in *Bimuria triquetra* that rim is well developed and distinctly undercut outwards.

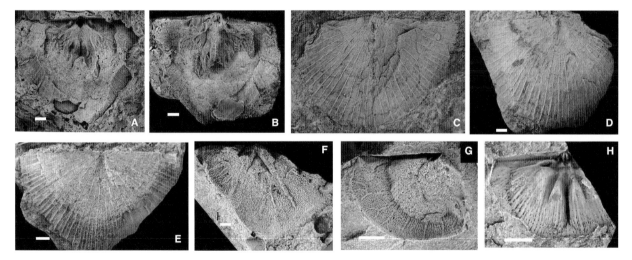

Fig. 20. **A–F.** *Acculina*? sp., Kopkurgan Formation (Katian), Locality 735; **A, B,** BC 63825a, dorsal internal mould and latex cast of it; **C,** BC 63818b, ventral exterior; **D,** BC 63833, dorsal exterior; **E,** BC 63871, dorsal exterior; **F,** BC 63887, ventral internal mould of juvenile individual. **G, H.** *Tesikella necopina* (Popov, 1980b); Kopkurgan Formation (late Sandbian), Locality 8235, **G,** BC 60853a, ventral internal mould, **H,** BC 60852, dorsal internal mould. Scale bars 2 mm.

Bimuria sp.

Plate 6, figures 21–23

Material. – Baigara Formation. Locality 1021: three articulated shells and two ventral valves including: NMW1256–1258, NMW98.28G.2034 (Pl. 6, fig. 21). Locality 1022: eight articulated shells, including NMW 98.28G.1945-1952. Locality 1023: five articulated shells, including 1963–1965, 1985 (Pl. 6, fig. 23). Locality 625/3, one articulated shell, NMW 98.28G.2035 (Pl. 6, fig. 22).

Remarks. – The shells show strong concentric ornament, and are mostly crowded with comae, almost orthocline ventral interarea with a narrow delthyrium partly covered by the pseudodeltidium. Dorsal interior observed on polished sections (Pl. 6, fig. 23) exhibits a thickened bema bordered by distinct rim, faint median ridge and two pairs of side septa placed close to each other, which is unusual for *Bimuria*. In addition to a well defined bema and two pairs of dorsal side septa, they differ from *Bimuria karatalensis*, which occurs upsequence in the same unit, in having a sagittal profile with the much less convex ventral valve and only gently concave dorsal valve.

Genus *Cooperea* Cocks & Rong, 1989

Type species (by original designation). – *Bimuria siphonata* Cooper, 1956, from the Pratt Ferry Formation (Darriwilian), Alabama, USA.

Cooperea sp.

Plate 6, figure 20

Material. – Baigara Formation, Locality 8121: ventral valve, BC 62400 (Pl. 6, fig. 20); dorsal external mould, BC 62401.

Remarks. – A smooth, strongly transverse ventral valve lacking a trail and comae are characteristic of *Cooperea*, and thus attribution to that genus seems probable for these specimens. They resemble *Cooperea aurita* Nikitin, Popov & Bassett, 2006, from the Angrensor Formation (Katian) of the Boshchekul Terrane in north-eastern Central

Kazakhstan, but without knowing the dorsal interior, the true specific identification remains uncertain.

Family Leptellinidae Ulrich & Cooper, 1936
Subfamily Leptellininae Ulrich & Cooper, 1936

Genus *Acculina* Misius *in* Misius & Ushatinskaya, 1977

Type species (by original designation). – *Acculina acculica* Misius *in* Misius & Ushatinskaya, 1977 (see below).

Acculina acculica Misius *in* Misius & Ushatinskaya, 1977

Plate 7, figures 1–12, 15

1977 *Acculina acculica* sp. nov. Misius; Misius & Ushatinskaya, p. 114, pl. 26, figs 21–24.

1985 *Acculina villosa* sp. nov. Nikitina, p. 24, pl. 1, figs 14–19.

1986 *Acculina acculica* Misius; Misius, p. 143, pl. 13, figs 3–30, pl. 14, figs 1–13.

Type specimens. – The holotype of *Acculina acculica* is Geological Institute, Bishkek, Kyrgyzstan 53/673, from the Tabylgaty Formation (Sandbian, *Nemagraptus gracilis* Biozone) of the Moldo-Too Range, north Kyrgyzstan. The holotype of *Acculina vilosa* is CNIGR 6/12093, from the Rgaity Formation (Darriwilian), Talapty Farm, south Kendykyas range, Kazakhstan.

Material. – Kopkurgan Formation, Locality 816: 31 conjoined valves, one ventral and one dorsal valves, including BC 62378 (Pl. 7, figs 2, 3), BC 60764–68. Locality 817: conjoined valves, BC 60769, dorsal valve, BC 60770. Baigara Formation, Locality 1021, NMW 98.28G.1129 (Pl. 7, fig. 1), 1130 (Pl. 7, fig. 4), NMW 98.28G.1131–1136, 2039-2053. Locality 1022: four conjoined valves, NMW 98.28G.1158–1160. Locality 625/3: a pair of conjoined valves, BC 60868. Locality 765: six pairs of conjoined valves, NMW 98.28G.1163-1168 (basic statistics for 13 articulated shells. Berkutsyur Formation, Locality 812: four

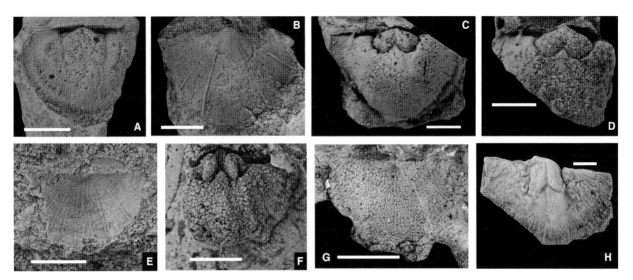

Fig. 21. **A–E.** *Mabella* sp. 1; **A,** Kopkurgan Formation (early Sandbian), Locality 735, BC 63846b, ventral internal mould. **B-D,** Berkutsyur Formation; **B,** Locality 815, BC 60840, ventral exterior; **C,** Locality 812, NMW 98.28G.2034, ventral internal mould; **D,** Locality 812, NMW 98.28G.2035, dorsal exterior. **E,** Kipchak Limestone (late Darriwilian – early Sandbian), Locality 166, BC 62442, dorsal view of articulated shell. **F, G.** *Leangella* (*Leangella*) sp.; Kopkurgan Formation (Katian), Locality 735; **F,** BC 63830e, ventral internal mould; **G,** BC 63830f, dorsal external mould. **H.** *Leptellina* sp. Baigara Formation (Darriwilian), Locality 1021, NMW 98.28G.1892, ventral internal mould. Scale bars 2 mm.

ventral and five dorsal valves, including BC 62322 (Pl. 7, fig. 9), BC 62323 (Pl. 7, fig. 12), BC 62324 (Pl. 7, fig. 10), BC 60874 (Pl. 7, fig. 11). Locality 815: three pairs of conjoined valves, BC 60872 (Lv, 14.1; Ld, 14.2; W, 18.6; T, 5.7), NMW 98.28G.1126 and 1127; and a dorsal internal mould, BC 62369 (Pl. 7, fig. 15). Rgaity Formation (North Tien Shan Terrane): Locality 12211, three ventral and two dorsal valves preserved as internal and external moulds, including NMW 98.66G.1161.1 (Pl. 7, figs 5, 8), 2117 (Pl. 7, figs 6, 7), 2118 and 2119. Average sizes for 20 articulated shells from Baigara Formation (localities 625, 1021 and 1022). Length: X, 8.7; S, 1.02; min, 6.8; max, 10.5. Width: X, 11.7; S, 2.03; min, 19.8; max, 19.6. Thickness: X, 2.3; S, 0.27; min, 1.9; max, 2.9.

Remarks. – Densely spaced comae covering the anterior half of the shell is an invariable characteristic of all known populations of *Acculina* from the Chu-Ili Terrane (Popov *et al.* 2002), as well as the shells described by Nikitina (1985) as *Acculina villosa* from the Rgaity Formation of the southern Kendyktas Range in the North Tien-Shan Terrane. However, the presence of comae was remarkably omitted in the previous descriptions of *Acculina acculica* by Misius (*in* Misius & Ushatinskaya 1977; Misius 1986), since most of the specimens illustrated in these publications are either exfoliated or preserved as internal moulds. Otherwise our studied specimens have no substantial differences from topotypes of *Acculina acculica* and are thus considered conspecific. Nikitina

(1985) pointed out that the presence of comae is the only difference between *Acculina villosa* and *Acculina acculica*, and therefore both taxa are considered here as a single species.

Acculina? sp.

Figure 20A–F

Material. – Kopkurgan Formation; Locality 735: one articulated shell, three ventral and three dorsal valves, including BC 63818b (L, 19.6 mm; W, 29.7; Fig. 20C), BC 63837f (L, 20.4; estimated W, 31), BC 63833 (L, 19.8; W, 32.0; Fig. 20D), BC 63871 (Fig. 20E), BC 63825a (Fig. 20A, B) and BC 63887 (Fig. 20F).

Remarks. – In addition to the many other plectambonitoids which have been found at Locality 735, there are seven very large specimens, mostly exceeding 30 mm in width, which clearly represent a different species to all the others. Ventral valve exteriors have normal convexity for about half their lengths and then become evenly resupinate antero-laterally without geniculation (Fig. 20C). The dorsal valve is almost flat posterior to the mid length with a faint, shallow sulcus originating at the umbo and fading posteriorly (Fig. 20E), then it becomes strongly resupinate anterior to the mid-length. The ornament is unequally parvicostellate (Fig. 20C, E). A single artificially prepared ventral internal mould (Fig. 20F) exhibits a relatively small, bilobed ventral muscle field with diductor scars touching each other in front of significantly shorter adductor scars. Well-impressed mantle canals are lemniscate with short, straight, widely divergent *vascula media*. The dorsal valve interior (Fig. 20A, B) has a robust, trifid cardinal process, strongly thickened and widely divergent socket plates and a massive median septum strongly raised and bifurcating anteriorly. Since *Acculina* is also gently resupinate, the specimens can only be provisionally identified as *Acculina*? sp., although in having a large, resupinate shell with strongly geniculate dorsal valve it recalls *Acculina phyaliformis* (Klenina, *in* Klenina *et al.* 1984), but differs in its smaller ventral muscle field with widely divergent diductor scars and a massive, anteriorly bifurcated dorsal median septum and a weakly defined rim bounding the lophophore platform. Until more material becomes available this taxon is left in open nomenclature; nevertheless, the specimens undoubtedly represent an undescribed species and possibly a new genus.

Genus *Dulankarella* Rukavishnikova, 1956

Type species (by original designation). – Dulankarella magna Rukavishnikova, 1956, from the Dulankara Formation (early Katian), Chu-Ili Range, Kazakhstan.

Dulankarella larga Popov, Cocks & Nikitin, 2002

Plate 7, figure 13; Plate 8, figures 1–5

2002 *Dulankarella larga* sp. nov. Popov, Cocks & Nikitin, p. 39, pl. 4, figs 9–25, pl. 5, figs 1–3.

Holotype. – BC 57421, dorsal valve interior, from the Anderken Formation (Sandbian) of Buldukbai-Akchoku, Chu-Ili Range, southern Kazakhstan.

Material. – Berkutsyur Formation, Locality 812: one dorsal external mould, BC 60875. Locality 813: one ventral valve, BC 60818 (Lv, 13.8; T, 4.9) and one dorsal external mould, BC 60842. Locality 815: one articulated shell, one ventral and two dorsal valves, including BC 62363 (L, 6.7 mm; West, 14 mm; Pl. 7, fig. 13); BC 62364 (Lest, 12; W, 12.8), BC 62365 and NMW 98.28G.2258. Locality 817: articulated shell, BC 62390 (Pl. 8, figs 1-3) and five ventral valves,

including BC 60791-3, BC 62388–89. Kopkurgan Formation, Locality 816: four articulated shells, five ventral and six dorsal valves, including BC 60817, BC 62376 (L, 11.4 mm; West, 18 mm; Pl. 8, fig. 5) BC 62377 (Lv, 17.2; W, 17.8; Pl. 8, fig. 4), BC 62381–85. Total, five articulated shells, 12 ventral and seven dorsal valves.

Remarks. – This species was described by Popov *et al.* (2002); however, although the ventral mantle canals of *Dulankarella larga* were originally identified as saccate, new well-preserved ventral internal moulds of this species (Pl. 8, fig. 5) clearly show that they are in fact lemniscate. Other species of the genus are *Dulankarella partita* Percival (1979) from the Goonumbla Volcanics (Katian) of Gunningbland, New South Wales, *D.* sp. from the Angrensor Formation (late Katian) of the Boshchekul Terrane (Nikitin *et al.*, 2006), and *D*? sp. from the Taldyboi Formation (early to mid Katian) of the Chingiz-Tarbagatai Terrane (Popov & Cocks 2014). While generic attribution of the shells, from the Marinikha Regional Stage (upper Katian) of the Sayany-Altai Region of Siberia, assigned by Kulkov & Severgina (1989) to *Dulankarella magna* is likely, the species cannot be confirmed due to inadequate illustrations in the cited paper. However, *D. namasensis* and its junior synonym *D. subquadrata*, both named by Klenina (*in* Klenina *et al.* 1984), and both from the Taldeboy Formation in the Chingiz-Tarbagatai Terrane should be re-assigned to *Shlyginia* (Nikitin & Popov, 1996; Popov & Cocks, 2014, p. 697).

Genus *Kajnaria* Nikitin & Popov *in* Klenina, Nikitin & Popov, 1984

Type species (by original designation). – Kajnaria derupta Nikitin & Popov *in* Klenina *et al.* 1984, see below.

Kajnaria derupta Nikitin & Popov *in* Klenina, Nikitin & Popov, 1984

Plate 8, figures 7–17; Figure 23E

1984 *Kajnaria derupta* sp. nov. Nikitin & Popov; Klenina, Nikitin & Popov, p. 145, pl. 18, figs 10, 13–16.

Holotype. – CNIGR 12095/108, upper part of Bestamak Formation (early Katian), Chingiz Range, Kazakhstan.

Material. – Baigara Formation, Locality 1026: four ventral internal moulds and one dorsal external mould plus ventral interarea, including NMW 98.28G.1169 (Pl. 8, figs 9, 10), BC 12931 (Pl. 8, fig. 11). Locality 1026b: eight articulated shells, NMW 98.28G.2002 (Pl. 8, figs 12, 13), NMW 98.28G.2000 (Pl. 8, figs 7, 8; Text-fig. 23E), 2001 (Pl. 8, figs 14–17), 2002–2006. Total nine articulated shells and four ventral valves.

Remarks. – Shells from the Baigara Formation have a strongly impressed ventral muscle field completely enclosed by strong muscle bounding ridges and a pair of curved ridges, which form the characteristic w-shaped structure diagnostic of *Kajnaria*. They can be assigned with confidence to *K. derupta*, which was originally described by Nikitin & Popov (*in* Klenina *et al.* 1984) from the Chingiz-Tarbagatai Terrane, where it occurs in the Bestamak Formation (early Sandbian). In addition to this new occurrence in the Chu-Ili Terrane, *K. derupta* was also reported by Gridina *et al.* (2004) from the Narulgen Formation (early Sandbian) of the Boshchekul Terrane.

Genus *Leptellina* Ulrich & Cooper, 1936

Type species (by original designation). – *Leptellina tennesseensis* Ulrich & Cooper, 1936, from the Arline Formation (Darriwilian) of Tennessee, USA.

Leptellina sp.

Plate 7, figures 16–18, 20–22; Figure 21H

Material. – Baigara Formation, Locality 1021: 58 articulated shells and two dorsal valves, including NMW 98.28G.1292 (Fig. 21H); 1892 (Pl. 7, figs 20, 21), 1893 (Pl. 7, fig. 22), 1894 (Pl. 7, fig. 18), 1895–1943. Locality 1022: eight articulated shells, including NMW 98.28G.1265–1272. Locality 1023: one articulated shell, NMW 98.28G.1944 (Pl. 7, figs 16, 17). Total 66 articulated shells, one ventral and two dorsal valves.

Description. – Shell concavo-convex, transverse, semioval, about 65-80% as long as wide with maximum width at the hinge line. Cardinal extremities acute, anterior commissure rectimarginate. Ventral valve lateral profile unevenly convex with maximum height at about one-third valve length from the anterior margin in mature individuals. Ventral interarea relatively high, almost orthocline, with a narrow, triangular delthyrium covered apically by small, convex

pseudodeltidium. Dorsal valve moderately concave with a low, planar catacline interarea bearing a small, open notothyrium, occluded by the cardinal process. Concentric ornament of dense comae on the ventral valve, interrupted by the accentuated ribs. Radial ornament parvicostellate with 5 primary ribs originated at the umbo and two generations of secondary costellae inclined in broad interspaces between them (Pl. 7, fig. 16), totalling up to 25 accentuated ribs along the margins of mature individuals (Pl. 7, fig. 21). Interspaces between ribs occupied by very fine parvicostellae about 8–10 per mm and about 8–11 interspaces between accentuated ribs. Ventral interior with small teeth supported by short, widely divergent dental plates. Muscle field strongly impressed, bilobed, about 80% as long as wide and 40% valve length. Diductor scars long, divergent, not completely enclosing significantly shorter, subtriangular adductor scar (Pl. 7, fig. 22). Mantle canals strongly impressed, lemniscate with short, divergent *vascula lateralia*. A distinct subperipheral rim may be present along the anterior and lateral valve margins. Dorsal interior with trifid cardinal process with a strong central ridge and reduced lateral components. Other details of cardinalia unknown. Dorsal median septum thick, gradually raised anteriorly to merge with low rim bounding the platform, about two-thirds as long as the valve (Pl. 7, fig. 18).

Remarks. – While these shells probably belong to a new species, they are left in open nomenclature here mainly because of inadequate data on the dorsal interior. They can be clearly distinguished from the other Kazakh species *Leptellina seletensis* Nikitin & Popov, 1983, from the Isobilnaya Formation (early Sandbian) of the Selety Terrane, *L.* aff. *L. seletensis* Nikitin & Popov, 1983, and *L. subquadrata* (Rukavishnikova, 1956), both from the Uzunbulak Formation (Darriwilian) of the Chu-Ili Terrane, in having strongly developed comae on ventral valve, strong accentuated costellae separated by broad interspaces containing up to 11 parvicostellae, a larger ventral muscle field with diductor scars strongly exceeding the adductor scar in length, and a weakly defined rim bounding the dorsal valve platform.

Genus *Mabella* Klenina *in* Klenina, Nikitin & Popov, 1984

Type species (by original designation). – *Leptellina (Mabella) semiovalis* Klenina, *in* Klenina, Nikitin & Popov, 1984, from the Taldyboi Formation (Katian), Chingiz Terrane, Kazakhstan.

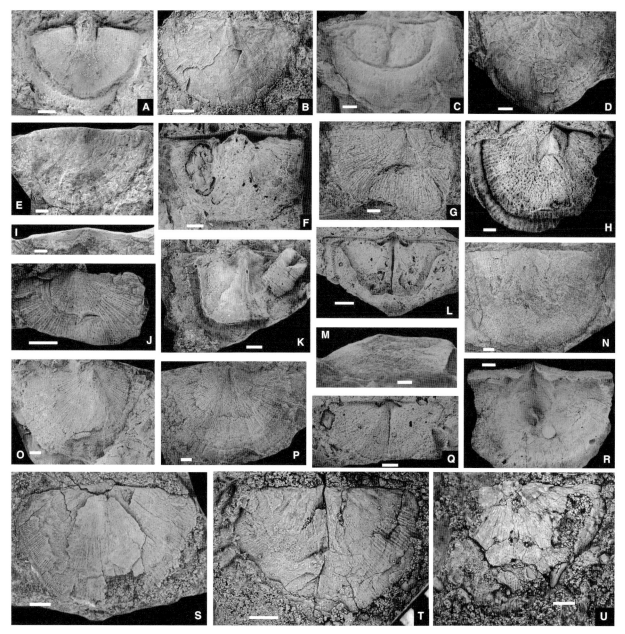

Fig. 22. **A–E, I.** *Chaganella chaganensis* Nikitin, 1974; Berkutsyur Formation, Locality 388; **A,** BC 63796, ventral internal mould; **B,** BC 63791, ventral exterior; **C,** BC 63794, dorsal internal mould; **D,** BC 63793, dorsal exterior; **E, I,** BC 63807, ventral exterior, posterior view of interarea; Scale bars are 2 mm. **F–H, K–R.** *Anoptambonites kovalevskii* Popov, Nikitin & Cocks, 2000; Kopkurgan Formation (Katian), Locality 735; **F,** BC 63838a, ventral internal mould; **G,** BC 63835b, dorsal valve; **H,** BC 63878, ventral internal mould; **K,** BC 63820a, latex cast of dorsal interior; **L,** BC 63842, dorsal internal mould; **M, N,** BC 63812, ventral exterior, side and ventral views; **O,** BC 63838d, latex cast of ventral exterior; **P,** BC 63841, dorsal exterior; **Q,** BC 63891a, dorsal internal mould; **R,** BC 63837a, latex cast of ventral exterior. **J.** *Anoptambonites* sp. 1; Kopkurgan Formation (Katian), Locality 727; BC 63843, ventral valve. **S–U.** *Anoptambonites* sp. 2; Unnamed formation (late Katian), Locality 8126; **S,** BC 63809, exfoliated ventral exterior; **T,** BC 63808, exfoliated dorsal exterior; **U,** BC 63810, exfoliated ventral exterior. Scale bars 2 mm.

Mabella sp.

Figure 21A–E

Material. – Berkutsyur Formation, Locality 812: five ventral and four dorsal valves, including BC 62334-

36, NMW 98.28G.2034 (Fig. 21C), NMW 98.28G.2035 (Fig. 21D). Locality 815: ventral valve, BC 60840 (Lv = 4.1, W = 5.5; Fig. 21B). Locality 817: ventral internal mould, BC 63846b (Fig. 21A). Takyrsu Formation (Kipchak Limestone), Locality 166: articulated shell, BC 62442 (Fig. 21E).

Remarks. – In size, proportions, and sagittal profile of the strongly convex ventral valve and strongly impressed, bilobed ventral muscle field, these shells closely resemble *Mabella conferta* (Popov, 1985) but differ in their strongly differentiated parvicostellate radial ornament with five accentuated ribs originating at the umbo and 10 to 12 along the shell margins of full grown individuals. In the absence of data on the dorsal interior further comparison is impossible, therefore the shells from the Berkutsyur Formation are left in open nomenclature although their generic attribution is in no doubt. *Mabella multicostata* (Rukavishnikova, 1956), from the Dulankara Formation (Katian) of the Chu-Ili Terrane, requires revision. Unlike *Mabella* sp. 1 of Popov & Cocks (2014), the Berkutsyur material is characterised by parvicostellate ornament with faint and numerous accentuated ribs, about 4 to 5 per 2 mm along the anterior margin and a somewhat larger shell. Klenina (*in* Klenina *et al.* 1984) described three species of *Mabella*, but they were synonymised with each other by Popov & Cocks (2014) when revising the Taldyboi Formation fauna from the Chingiz-Tarbagatai Terrane.

Subfamily Palaeostrophomeninae Cocks & Rong, 1989

Genus *Apatomorpha* Cooper, 1956

Type species (by original designation). – *Rafinesquina pulchella* Raymond, 1928, from the Athens Formation (Darriwilian) of Tennessee, USA.

Apatomorpha akbakaiensis n. sp.

Plate 8, figures 18–22; Plate 9, figures 1–4

Derivation of name. – After the Akbakai gold mine situated south-east of the type locality.

Holotype. – NMW 98.28G.1240 (L = 42.6, W = 38.0, T = 8.0; (Pl. 9, figs 1–3), articulated shell, from Locality 1022, Baigara Formation (late Darriwilian to early Sandbian), area *c.* 7 km SW of Baigara Mountain, south Betpak-Dala, South Kazakhstan.

Paratypes. – Baigara Formation, Locality 1022: 9 articulated shells, including NMW 98.28G.1230–1239 (L, 32.6; W, 34.0; T, 9.4; Pl. 8, figs 18–20); Locality 1023: 20 conjoined valves, including NMW 98.28G.1241 (Pl. 8, fig. W), 1242 (Pl. 8, fig. 21), 1243 (Pl. 9, fig. 4), NMW 98.28G.1244–1259.

Diagnosis. – Shell large for the genus, flatly concavoconvex, subquadrate outline, high interareas of both valves about equal height, radial ornament finely and unequally parvicostellate.

Description. – Shell large, flatly concavoconvex, outline subquadrate, almost as long as wide with maximum width at or slightly anterior of the hinge line and 20–25% as thick as long. Cardinal extremities almost right angled, anterior commissure rectimarginate. Lateral shell margins subparallel, almost straight, anterior margin broadly rounded. Ventral valve profile unevenly convex, with maximum height at about one-quarter valve length. Umbonal area gently convex, may be weakly carinate. Ventral interarea high, planar almost orthocline to slightly apsacline with a very narrow subtriangular delthyrium covered by the convex pseudodeltidium. Dorsal valve gently and unevenly concave with maximum curvature in the umbonal area. Dorsal umbonal area with a shallow subcircular depression extending anteriorly as a shallow sulcus fading completely near mid-length. Dorsal interarea almost as high as the ventral pseudointerarea, planar, hypercline with a narrow, convex chilidium completely covering the nothothyrium (Pl. 9, fig. 1). Radial ornament finely and unequally parvicostellate, with 11 strong ribs originating at the umbo and numerous accentuated ribs, about 2 per mm, at the anterior valve margin. Interspaces between ribs covered by 5–7 very fine parvicostellae. Concentric ornament of very fine, densely-spaced fila. Ventral interior with strong teeth and short dental plates. Ventral muscle field strongly impressed, flabellate, bounded laterally by straight muscle bounding ridges gradually fading anteriorly. Adductor muscle scar wide, subtriangular, about twice as short as narrow, widely divergent diductor scars. Ventral mantle canals saccate with very short *vascula media* branching in front of the adductor scars (Pl. 8, figs 21, 22). Dorsal interior (Pl. 9, fig. 4) with a prominent, blade-like cardinal process, high, widely divergent socket plates and a low, thick median septum terminating at about one-quarter valve length from the anterior margin.

Remarks. – *Apatomorpha akbakaiensis* differs from *A. pulchella* (Raymond, 1928), from the Athens Formation (Sandbian) of Tennessee, in being more than twice as large, as well as in having finely parvicostellate (not multicostellate) radial ornament, a high dorsal interarea, about equal height with the ventral interarea, and a less transverse, almost subquadrate shell outline. Two other species of the genus are *Apatomorpha cita* (Rukavishnikova, 1956), from the Uzunbulak Formation (Darriwilian) of the Chu-Ili Terrane and *A. melrosensis* Laurie, 1991, from the

Cashions Creek Limestone (Sandbian) of Australia. Like *Apatomorpha akbakaiensis*, both have parvicostellate radial ornament; however, the new species differs from them in having a large, flatly concavoconvex shell of subquadrate outline with high interareas of about equal size in both valves. The new species also differs from *Apatomorpha cita* in the absence of dorsal valve geniculation. As noted by Cocks & Rong (1989), *Apatomorpha altaicus* [sic] Severgina, 1960, does not belong to the genus because it has a platform, and that species probably belongs to *Toquimia*.

Genus *Ishimia* Nikitin, 1974

Type species (by original designation). – *Ishimia ishimensis* Nikitin, 1974, see below.

Ishimia aff. *I. ishimensis* Nikitin, 1974

Plate 9, figures 5–10, 13

Material. – Berkutsyur Formation, Locality 812: three articulated shells, one ventral and five dorsal valves. Locality 813, 11 articulated shell five ventral and five dorsal valves, including; ventral internal moulds, BC 62352 (Lv, 11; W, 18.0; Pl. 9, fig. 13), BC 62353 (Lv, 15.8; West, 24; Pl. 9, fig. 10), and BC 62354 (Lv, 11.0; West, 16). Locality 814: ventral external mould, BC 60835. Locality 815: four articulated shells, three ventral and three dorsal valves, including BC 62368, BC 60785-88. Locality 8120: three articulated shells, two ventral and two dorsal valves, including BC 60834, BC 60836–9. Locality 8124, nine articulated shells, five ventral and two dorsal valves, including BC 60732–41, BC 62416 (Pl. 9, fig. 7), BC 62417 (Pl. 9, fig. 6), BC 62418 (Pl. 9, fig. 8), NMW 98.28G.1984 (Pl. 9, fig. 9). Total 33 articulated shells, 17 ventral and 17 dorsal valves.

Remarks. – Shells from the Berkutsyur Formation are very similar to the types of *Ishimia mediasiatica*, which occur in the Tabylgaty Formation (*Ishimia* Beds, *Nemagraptus gracilis* Biozone) of the North Tien Shan Terrane and *Ishimia ishimensis* Nikitin, 1974, from the Kreshchenovka Formation (early Sandbian) of the Kokchetav-Kalmykkol Terrane in having distinct unequally parvicostellate radial ornament, shell size and proportions, and especially in the characters of the ventral muscle field and mantle canals. The main difference from *Ishimia mediasiatica* is in the parvicostellate radial ornament, which shows strong differentiation in size between accentuated ribs and parvicostellae. Also, the accentuated ribs in *Ishimia* aff. *I. ishimensis* are less numerous and separated by wider interspaces occupied by up to 8 parvicostellae (Pl. 9, figs 5–7), while in *Ishimia mediasiatica* their number is no more than 3. In comparison with topotypes of *Ishimia ishimensis*, the shells from the Berkutsyur Formation are considerably smaller and have a distinct subperipheral rim in the ventral valve (Pl. 9, figs 10, 13), which is also present in *Ishimia mediasiatica* but absent in *Ishimia ishimensis*. It is possible that all these shells should be assigned to a single morphologically variable species, but that can be resolved only after revision of the type material. *Ishimia* aff. *I. ishimensis* occupies an intermediate position in shell morphology between *Ishimia mediasiatica* and *Ishimia ishimensis*, and thus its species identification is provisional.

Genus *Lepidomena* Laurie, 1991

Type species (by original designation). – *Lepidomena pulchra* Laurie, 1991, from the Benjamin Limestone (Sandbian) of Tasmania, Australia.

Lepidomena betpakdalensis n. sp.

Plate 10, figures 1–9

Derivation of name. – After the Betpak-Dala desert.

Holotype. – NMW98.28G. 1324 (Pl. 10, figs 4–6), articulated shell, articulated shell, from Locality 1022, Baigara Formation (late Darriwilian to early Sandbian), area *c.* 7 km south-west of Baigara Mountain, south Betpak-Dala.

Material. – Locality 1022: 52 articulated shells, including paratypes NMW 98.28G.1323 (Pl. 10, figs 2, 3), 1325 (Pl. 10, figs 7, 8), 1326 (Pl. 10, fig. 9), 1327–1357, 1376, 1377. Locality 1023: 42 articulated shells and one ventral valve, including paratypes, NMW 98.28G.1289 (Pl. 10, fig. 1), NMW 98.28G.1290–1322, 1360-1368. Locality 625/1 (coll.

T.B. Rukavishnikova): six articulated shells, NMW 98.28G.1369–1371, 1373–1375. Locality 625/3 (coll. T.B. Rukavishnikova): 16 articulated shells, NMW 98.28G.1274–1288, 1372. Average sizes for 29 articulated shells from Locality1022. Length: X, 16.6; S, 3.00; min, 11.0; max, 23.8. Width: X, 29.7; S, 4.44; min, 20.0; max, 38.1. Thickness: X, 6.75; S, 1,72; min, 3.5; max, 10.0

Diagnosis. – Shell of medium size for the genus with strong dorsal geniculation of both valves, unequally parvicostellate radial ornament and a massive, lens-like cardinal process.

Description. – Shell concavoconvex, geniculated anteriorly, transverse, semioval in outline on average 56.6% (S, 8.01; OR, 39.6–78%; N, 29) with maximum width at the hinge line and 42.5% (S, 14.4, OR, 17.7–74.1) as thick as long. Cardinal extremities acute and slightly alate. Anterior commissure rectimarginate. Ventral valve sagittal profile gently convex posterior to geniculation, with maximum height at the geniculation, about one-third valve length in adult individuals. Ventral pseudointerarea low, planar, orthocline to slightly apsacline with a narrow, triangular delthyrium completely covered by the convex pseuododeltidium which has a distinct median ridge. Dorsal valve gently convex to almost flat anterior to the geniculation, with a low, planar, apsacline interarea two-thirds shorter than the ventral interarea. Narrow notothyrium completely covered by a convex chilidium with a median ridge. Shallow dorsal sulcus originating at the umbo, gradually fading anteriorly to disappear in front of the geniculation. Radial ornament finely and unequally parvicostellate with 7–9 accentuated ribs originating at the umbo and two generations of secondary ribs between them. Accentuated ribs separated by interspaces up to 3 mm wide along the geniculation, within which are 12–14 parvicostellae with 10–14 per 3 mm (numbering 10–16 in 5, 7, 6, 3, 3, 1, 2 specimens). Anterior to the geniculation the number of parvicostellae in the sectors decreases to five or six due to a second generation of accentuated ribs. Concentric ornament with faint, densely spaced filae, 7–8 per mm. Ventral interior with strong teeth lacking dental plates and a flabellate muscle field bounded laterally by faint, straight widely-divergent muscle bounding ridges (Pl. 10, fig. 1). Adductor scar short, wide, subtriangular not enclosed by narrow, divergent diductor scars almost twice as long. Dorsal valve interior with high lens-like cardinal process (Pl. 10, fig. 9), high, short, widely-divergent socket plates,

wide and low median septum anteriorly merging with a rim bordering the platform.

Remarks. – *Lepidomena* has only been identified previously from Australia where there are two species, *Lepidomena pulchra* Laurie, 1991 and *L. fortimuscula* Laurie, 1991, both from the late Darriwilian to early Sandbian of Tasmania. The new species can be distinguished from both in having unequally parvicostellate radial ornament and strong dorsal geniculation of both valves. In addition, the massive cardinal process in the Kazakh shells is even stronger than those of the Australian species.

Genus *Tesikella* Popov, Cocks & Nikitin, 2002

Type species (by original designation). – *Palaeostrophomena necopina* Popov, 1980, from the Anderken Formation (Sandbian) of the Chu-Ili Range, southern Kazakhstan.

Tesikella necopina (Popov, 1980b)

Plate 10, figure 12; Figure 20G, H

1980b *Palaeostrophomena necopina* sp. nov. Popov, p. 145, pl. 1, figs 8–11.

2002 *Tesikella necopina* (Popov); Popov, Cocks & Nikitin, p. 46, pl. 5, figs 30, 31, pl. 6, figs 1–6.

Holotype. – CNIGR 15/11523, dorsal internal mould from the Anderken Formation (Sandbian), Locality 127 on the east side of Kopalysai, Chu-Ili Range, southern Kazakhstan.

Material. – Berkutsyur Formation, Loc 8235: one articulated shell, two ventral and one dorsal valves, including BC 62436 (L, 7.0 mm; W. 14.2 mm; Pl. 10, fig. 12), BC 60852 (Fig. 20H), BC 60853a (Fig. 20G), BC 60853b, c.

Remarks. – *Tesikella* has only been previously identified from its type species in the Anderken Formation. It was confined to the shallow marine environment nearshore which was later populated by brachiopods of the rhynchonellide biofacies. Yet a few specimens of *Tesikella* were found in graded siliciclastic rocks at the base of the Kopkurgan Formation which were deposited in a basinal environment after they had been transported downslope in a mass flow. The shells in our collection are mostly broken, yet it is possible to see characteristic features of the interiors which are diagnostic of *Tesikella*, including strong double teeth lacking dental plates, a large ventral muscle field enclosed by bilobed muscle bounding ridges, a well defined ventral subperipheral rim, and a well developed dorsal platform, all of which leave no doubts about the generic attribution.

Genus *Titanambonites* Cooper, 1956

Type species (by original designation). – *Titanambonites medius* Cooper, 1956, from the Athens Formation (Sandbian) of Tennessee, USA.

Titanambonites cf. *T. magnus* Nikitin, 1974

Plate 9, figures 11, 12, 14, 15

cf. 1974 *Titanambonites magnus* sp. nov. Nikitin, p. 56, pl. 5, figs 1–5.

Material. – Baigara Formation, Locality 1026: ventral internal mould, BC 62447 (Pl. 9, fig. 12). Locality 1026b: two internal moulds of articulated shells, NMW 98.28G.1986, 1987 (Pl. 9, figs 11, 14); NMW 98.28G.2109 (Pl. 9, fig. 15).

Remarks. – While only internal moulds of a few specimens are in the collection, they exhibit characteristic features suggesting close affinity to *Titanambonites magnus* Nikitin, 1974, from the Kreshchenovka Formation (early Sandbian) of the Kokchetav-Kalmykkol Terrane, including a transverse, strongly concavoconvex shell, a large, flabellate ventral muscle field extending anteriorly almost to the mid-valve and bounded laterally by strong, straight, anteriorly divergent muscle bounding ridges, the absence of a platform in the dorsal valve, as well as in the size and proportions of the shell. However, because of the absence of data on the exterior, the taxonomic identification remains provisional.

Genus *Shlyginia* Nikitin & Popov, 1983

Type species (by original designation). – *Shlyginia declivis* Nikitin & Popov, 1983, from the

Kreshchenovka Formation (early Sandbian) of the Kalmykkol–Kokchetav Terrane, north-central Kazakhstan.

Remarks. – In addition to the type species, the other *Shlyginia* previously identified from Kazakhstan are *S. extraordinaria* (Rukavishnikova, 1956) from the Dulankara Formation, which was revised by Popov *et al.* (2000) and Popov & Cocks (2006), *S. fragilis* (Rukavishnikova, 1956) from the Anderken Formation, which was revised by Popov *et al.* (2002), *S. perplexa* Nikitin & Popov (1996) from the Betpak-Dala mudmound, *S. solida* described by Nikitin & Popov *in* Klenina *et al.* (1984) from the Erkebidaik Formation, *S. namasensis* (Klenina *in* Klenina *et al.*, 1984) from the Taldyboi Formation, as well as the *Shlyginia*? sp. described by Nikitin *et al.* (2006) from the Angrensor Formation, and *Shlyginia* sp. 1 and *Shlyginia* sp. 2 from the Akdombak Formation (Popov & Cocks, 2014).

Shlyginia extraordinaria (Rukavishnikova, 1956)

Plate 10, figures 13–16

1956 *Dulankarella extraordinaria* sp. nov. Rukavishnikova, p. 138, pl. 3, figs 1–3.

2006 *Shlyginia extraordinaria* (Rukavishnikova); Popov & Cocks, p. 268, pl. 4, figs 22–23, 25–26 (full synonymy).

Holotype. – IGNA 35/1369, ventral internal mould; Dulankara Formation, Degeres Member (mid Katian), Raisai, south Chu-Ili Range.

Material. – Locality 1501, nine ventral and one dorsal valves preserved as internal and external moulds, including NMW 98.28G.1960 (Pl. 10, fig. 15), 1961 (Pl. 10, figs 13, 14), 1962.1 (Pl. 10, fig. 16), 2070. NMW 98.28G.1963–8.

Remarks. – The specimens from an unnamed formation at Ergenekty show a characteristic sagittal profile of both valves with distinct dorsal geniculation and a subperipheral rim along the margin of the ventral valve which were considered indicative of *Shlyginia extraordinaria* by Popov *et al.* (2000), and, in spite of distortion of the shells, there is no doubt about their species assignment. The taxon is common in the Otar and Degeres Beds of the Dulankara Formation in the South Chu-Ili Range, but is unknown from the other Kazakh terranes.

Shlyginia sp.

Plate 8, figure 6; Figure 24I

Material. – Takyrsu Formation, Locality 166: one ventral valve, BC 62445 (Lest, 8 mm; West, 12 mm). Locality 8135, one ventral internal mould NMW 98.28G.2163 (Pl. 8, fig. 6) and one dorsal external mould 2164, (Fig. 24I).

Remarks. – While *Shlyginia* is one of the most common taxa in the Anderken Formation (late Sandbian) of Chu-Ili, it is almost absent in coeval deposits in the West Balkhash region. Nevertheless these two shells are from graded siliciclastic rocks at Locality 8135 and were probably displaced from their original habitat within a mass flow. The specimens show finely parvicostellate radial ornament and ventral interiors characteristic of the genus, although their species attribution is problematic due to lack of adequate material.

Family Leptestiidae Öpik, 1933

Genus *Leangella* (*Leangella*) Öpik, 1933

Type species (by original designation). – *Leangella triangularis* Öpik, 1933, a junior synonym of *Leptaena scissa* J. de C. Sowerby, 1839, from the Upper Haverford Mudstone Formation (Rhuddanian) of Haverfordwest, Pembrokeshire, Wales (Cocks 1970).

Leangella (*Leangella*) sp.

Figures 19L; 21F, G

Material. – Kopkurgan Formation, Locality 735: eight ventral and three dorsal valves, including BC 63824d (Fig. 19L), BC 63827a, BC 63830e (Fig. 21F); BC 63830f (Fig. 21G), BC 63847b, BC 63888b, and BC 63891b.

Remarks. – The faint parvicostellate external ornament is poorly preserved in the available specimens (e.g. Fig. 21G), yet five accentuated ribs are visible. Interior of both valves shows the characteristic features of *Leangella* (*Leangella*) including a strongly impressed, bilobed ventral muscle field with diductor scars almost touching each other in front of shorter adductor scars, while the dorsal valve (Fig. 19L) has a bilobed, elevated

bema undercut anteriorly and a weak peripheral rim along the valve margin. Among the Kazakh species of the genus, the specimens from the Kopkurgan Formation seem closest to *Leangella* (*Leangella*) *bakanasensis* Popov & Cocks, 2014, from the Akdombak Formation (Katian) of the Chingiz-Tarbagatai Terrane, but they lack a distinct row of septules between the bema and the cardinal extremities and therefore they are probably not conspecific. They differ from *Leangella* (*Leangella*) *paletsae,* which occurs in the Degeres Beds (mid Katian) of the Chu-Ili Terrane, mainly in the absence of a dorsal median ridge. However, specific attribution of the shells from the Kopkurgan Formation remains uncertain due to their poorly preserved exteriors and insufficient number of specimens.

Family Xenambonitidae Cooper, 1956

Subfamily Aegiromeninae Havlíček, 1961

Genus *Akadyria* Nikitina, Neuman, Popov & Bassett *in* Nikitina *et al.,* 2006

Type species (by original designation). – *Akadyria simplex* (see below).

Akadyria simplex Nikitina, Neuman, Popov & Bassett *in* Nikitina *et al.,* 2006

Plate 11, figures 11, 13, 14

2006 *Akadyria simplex* Nikitina, Neuman, Popov & Bassett; Nikitina, Popov, Neuman, Bassett & Holmer, p. 188, figs 27.6–12.

Holotype. – USNM 485163, ventral internal mould, from the Uzunbulak Formation (Darriwilian) of Kurzhaksai, Chu-Ili Range, Kazakhstan.

Material. – Tastau Formation, Locality 154: disarticulated shell NMW 98.28G.1145 (Pl. 11, fig. 11); dorsal internal mould with ventral interarea, NMW 98.28G.1146 (Pl. 11, figs 13, 14).

Remarks. – *Akadyria simplex*, the only yet known species of the genus, is among the earliest representatives of the Aegiromeninae. It is a small, strongly flattened, concavoconvex to almost planoconvex shell, and has very simple shell morphology. The ventral valve has small teeth lacking dental plates and a weakly impressed ventral muscle field, while the internal shell surface is covered by concentric rows of faint papillae. The dorsal valve interior has no distinct features other than delicate cardinalia with undercut cardinal process and small, transverse socket plates. All these characters are present in the specimens from the Karakan Ridge, and they can be assigned to *Akadyria simplex* with confidence. At its type locality *Akadyria simplex* occurs in graptolitic argillites in association with linguliform brachiopods (Nikitina *et al.* 2006, p. 149), while at the Karakan Ridge locality it occurs within pelagic layers of distal turbidites deposited on the passive margin of the Chu-Ili Terrane facing the North-Tien Shan Microplate. It seems that the species was confined to the basinal environment of BA6 and was a strophomenide pioneer in the colonisation of the bathyal environment.

Genus *Chonetoidea* Jones, 1928

Type species (by original designation). – *Plectambonites papillosa* Reed, 1905, from the Slade and Redhill Mudstone Formation (Katian), Pembrokeshire, Wales.

Chonetoidea? sp.

Plate 11, figures 7–10

Material. – Berkutsyur Formation. Locality 8122: ventral external mould, BC 62407 (L, 3.1, W, 5.6; Pl. 11, fig. 8). Locality 8236, partly exfoliated exterior of conjoined valves, BC 62439 (L, 2.4; W, 4.2; Pl. 11, fig. 10); dorsal external mould of conjoined valves, BC 62441 (L, 3.5, W, 5.5, Pl. 11, fig. 9); dorsal internal mould, BC 62440 (L, 2.3; W, 3.7); dorsal external mould, BC 62437 (L, 3.5, W, 6.3; Pl. 11, fig. 7).

Remarks. – Only a few external moulds of this taxon are available and they are characterised by strongly transverse, flatly planoconvex to slightly concavoconvex shells with unequally parvicostellate radial ornament. Nine strong accentuated ribs originate at the umbo and there are up to 22 accentuated ribs along the anterior and lateral margins of the largest individuals. Wide interspaces between ribs are occupied by 6–9 faint parvicostellae. In major aspects of the external shell morphology these shells are indistinguishable from the *Chonetoidea?* sp. described by Nikitina *et al.* (2006) from the Uzunbulak Formation (Darriwilian) of the southern Chu-Ili Range, and are probably conspecific, but in the absence of

data on the interiors, the species present in the Kopkurgan Formation cannot be named.

Genus *Kassinella* Borissiak, 1956

Type species (by original designation). – *Kassinella globosa* Borissiak, 1956, from Katian beds on the Dulygaly-Zhilchik River, Ulutau, North Tien-Shan Terrane, Kazakhstan.

Remarks. – *Kassinella* was widespread in the Late Ordovician, with species identified from Laurentia, South China, and various parts of Gondwana, as well as in several of the Kazakh terranes (Cocks & Rong 1989). From the Chingiz-Tarbagatai Terrane, Popov & Cocks (2014) revised the type species *Kassinella globosa* Borissiak, as well as *K. tschingisensis* Klenina, 1984, and erected *K. kazbalensis*. From the Chu-Ili Terrane, Popov *et al.* (2002) described *Kassinella?* sp. from the Late Sandbian Anderken Formation

Kassinella simorini n. sp.

Plate 11, figures 1–6

1960   *Chonetoidea simorini* Borissiak (*nomen nudum*) Sokolskaya, pl. 27, figs 24, 25.

2002   *Kassinella* sp. Popov Cocks & Nikitin, p. 48, pl. 3, fig. 24.

Derivation of name. – In memory of A. M. Simorin, pioneering research worker on the Palaeozoic geology and brachiopods of Kazakhstan, a victim of Stalin's purges.

Holotype. – NMW 98.28G.1983 (Pl. 11, figs 1–4), disarticulated dorsal and ventral valve external and internal moulds; Locality 8235, Kopkurgan Formation (early Sandbian), about 7 km south-west of Lake Alakol, West Balkhash.

Paratypes. – Anderken Formation, Locality 538: three ventral valves, including BC 63717 (L, 3.9, W, 6.0), BC 63718 (L, 4.7 mm; W 6.2), BC 63719 (L, 5.0; W, 6.8). Kopkurgan Formation, Locality 8235: one ventral valve, NMW 98.28G.1980 (Pl. 11, figs 5, 6), and three dorsal valves, including BC 60815-16, BC 63701 (L, 3.4; W 5.8). Total: one pair of disarticulated valves, four ventral, and three dorsal valves.

Diagnosis. – Small for the genus; radial ornament with a single generation of seven accentuated costae and up to 8 parvicostellae between them. Ventral muscle field gently impressed; ventral posterolateral swellings weakly developed; ventral subperipheral rim formed by a concentric row of tubercles occasionally present. Dorsal interior with straight, blade-like socket ridges free anterolaterally. Dorsal platform with undercut anterior and lateral margins. No dorsal subperipheral rim.

Description. – Shell concavoconvex, transverse semioval outline, from two-thirds to three-quarters as long as wide with maximum width at or slightly anterior to the hinge line. Cardinal extremities varying from slightly acute to slightly obtuse. Anterior commissure rectimarginate. Ventral valve strongly convex with maximum depth slightly posterior to mid-length, subcarinate posteriorly. Ventral interarea low, planar, with a small apical pseudodeltidium covering the narrow triangular delthyrium (Pl. 11, fig. 6). Dorsal valve gently convex with a low, planar, hypercline interarea and shallow sulcus originating at the umbo and fading completely at mid-length. Radial ornament of seven accentuated ribs originated at the umbo and up to 8 parvicostellae between them separated by narrow interspaces and increasing in number by intercalation; up to 12 parvicostellae per mm at the anterior margin (Pl. 11, fig. 1). Ventral valve interior with small transverse teeth lacking dental plates (Pl. 11, figs 3, 6). Ventral muscle field weakly impressed, bilobed, bounded laterally by faint muscle bounding ridges. Ventral adductor scars small, lanceolate, bisected medially by a myophragm and completely enclosed anteriorly by the longer diductor scars (Pl. 11, figs 5, 6). A weakly-defined subperipheral rim formed by the concentric row of tubercles present in some specimens. A pair of swellings in the ventral valve posterolateral corners, characteristic of most *Kassinella* species, is weakly prominent. Dorsal interior with undercut cardinal process and blade-like, straight, widely divergent socket plates free anterolaterally. Low, blade-like, subtriangular median septum originating at some distance from the posterior margin and merged anteriorly with a rim bounding a subcircular platform (Pl. 11, figs 3, 4). Adductor muscle scars not clearly impressed.

Remarks. – While *Kassinella* is widespread in offshore biofacies across most of the other Kazakh Terranes (Popov & Cocks 2017), there are only few reports of its occurrence in Chu-Ili. The *Kassinella* sp. of Popov *et al.* (2002) from the upper part of the Anderken Formation is probably attributable to

Kassinella simorini, while a single ventral internal mould of *Kassinella* illustrated and briefly described by Nikitin *et al.* (1980) from the uppermost part of the Chokpar Formation is probably the latest known representative of the genus. Three other species of *Kassinella* are known from Kazakhstan, *K. globosa* Borissiak, 1956, from the Dulygaly and Ichkebash formations (early Katian) of the North Tien Shan Terrane; *K. kasbalensis* Popov & Cocks, 2014, from the Akdombak Formation (late Katian), and *K. tchingisensis* Klenina *in* Klenina *et al.*, 1984, from the Taldyboi Formation (early Katian) both in the Chingiz-Tarbagatai Terrane, and the last species seems to be closest to the new one. *Kassinella simorini* can be distinguished from all the others in having less numerous accentuated ribs, not increasing in number with shell growth, weakly impressed ventral muscle field, poorly defined posterolateral swellings in the ventral valve, and an undercut border of the dorsal platform not forming a high lamellose rim. *Kassinella kasbalensis* is also significantly larger and has a strong dorsal subperipheral rim, neither characteristic for *Kassinella simorini*. This taxon was first illustrated in the Russian 'Osnovy Paleontologii' by Sokolskaya (1960) as *Chonetoidea simorini* (*nomen nudum*), and the name was given by M. A. Borissiak. While its authorship is not protected since, according to the rules of ICZN (International Commission of Zoological Nomenclature 1999, Article 13.1.1.), it was not supported by a description or definition of the taxon, nevertheless we preserve the name in respect for the worker who first recognised it. *Kassinella simorini* is closely similar to *Durranella rugosa* Percival, 1979 from the Malongulli Formation and *Durranella septata* Percival, 1979 from the Goonumbla Volcanics, both of Katian age in New South Wales, Australia, and Cocks & Rong (1989) placed both those species within *Kassinella*. *Kassinella simorini* differs from those two species mainly in the radial ornament of only seven accentuated ribs originating at the umbo with wider segments between them occupied by up to eight parvicostellae, and a ventral subperipheral rim formed by tubercles. In addition, it differs from *Kassinella rugosa* in having a more pronounced ventral median ridge and in the complete absence of rugellae.

Genus *Synambonites* Zhan & Rong, 1995

Type species (by original designation). – *Synambonites biconvexus* Zhan & Rong, 1995 from the Xiazhen Formation (Katian), Zhejiang Province, China.

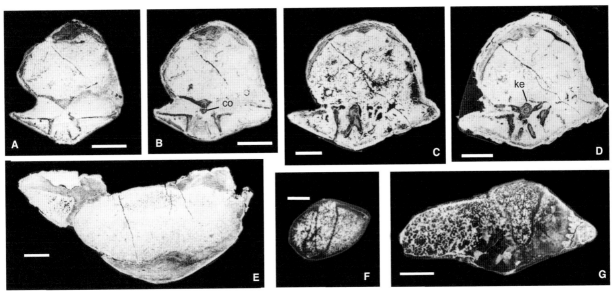

Fig. 23. **A–D.** *Grammoplecia globosa* (Nikitin & Popov, 1985); Baigara Formation (late Darriwilian – early Sandbian), Locality 1026b; NMW 98.28G.1462, transverse sections of a pair of conjoined valves at 1.7 mm, 1.9 mm, 2.0 mm and 2.7 mm from ventral umbo, abbreviations: co – cowl, ke – keel. **E.** *Kajnaria derupta* Nikitin & Popov *in* Klenina *et al.* 1984; Baigara Formation (late Darriwilian – early Sandbian), Locality 1026b, NMW 98.28G.2000, acetate replica of sagittal section of conjoined valves photographed in transmitted light. **F, G.** *Plectocamara* sp., Baigara Formation (late Darriwilian – early Sandbian), southern Betpak-Dala, Locality 765-e, NMW 98.28G.1982, acetate replica photographed in transmitted light; **F,** transverse section of articulated shell at 0.5 mm from ventral umbo; **G,** transverse section of distorted articulated shell showing spondylium simplex and discrete inner hinge plates. Scale bars 2 mm.

Synambonites sp.

Plate 13, figure 1

Remarks. – The only species of *Synambonites* hitherto described from Kazakhstan is *S. ricinium* Nikitin *et al.* (2006) from the early Katian Angrensor Formation in the Boshchekul Terrane. The single known specimen from Chu-Ili, a dorsal valve, BC 62404 (Ld, 3.5; W, 6.3) from the early Sandbian Berkutsyur Formation at Locality 8121, shows only the exterior of the dorsal valve: nevertheless it is well preserved and can be positively identified as *Synambonites*, It is thus the earliest known representative of the genus.

Genus *Tenuimena* Nikitina, Neuman, Popov & Bassett *in* Nikitina *et al.* 2006

Type species (by original designation). – *Tenuimena planissima*, see below.

Tenuimena aff. *T. planissima* Nikitina, Neuman, Popov & Bassett *in* Nikitina *et al.* 2006

Plate 10, figures 10, 11

aff. 2006 *Tenuimena planissima* Nikitina, Neuman, Popov & Bassett *in* Nikitina, Popov, Neuman, Bassett & Holmer, p. 188, figs 25.19–20, 27.1–3, 5, 28.B–C

Material. – Berkutsyur Formation, Locality 8122: four variably disarticulated shells, three ventral and four dorsal valves, including BC 62405a-b (Pl. 10, fig. 10), BC 62406 (Pl. 10, fig. 11), BC 62408b, BC 62408a, BC60771–81.

Remarks. – *Tenuimena planissima* is the only known species of the genus and is of Darriwilian age, in contrast to the Sandbian age of the new material. The latter might be within the morphological range of the former, but without better material, particularly of the ventral interior, that cannot yet be definitely confirmed.

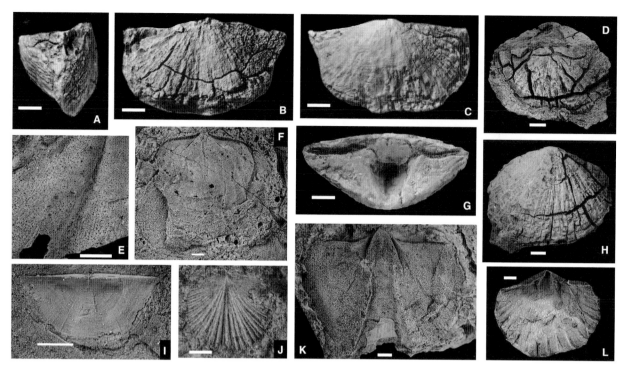

Fig. 24. **A–D, G, H.** *Atelelasma* sp.; Baigara Formation (late Darriwilian – early Sandbian), Locality 1026b; **A–C, G,** NMW 98.28G.2037, lateral, dorsal and ventral views of conjoined valves. **D, H,** NMW 98.28G.2038, dorsal and ventral views of conjoined valves. **E, F, K.** *Grammoplecia* sp., Locality 735; **E,** BC 63819a, external mould of dorsal exterior showing faint radial capillae; **F,** BC 63826, ventral internal mould; **K,** BC 63885a, ventral internal mould. **I,** *Shlyginia* sp.; Kopkurgan Formation (early Sandbian), Locality 8235; NMW 98.28G.2164, latex cast of dorsal exterior. **J,** Hesperorthidae gen. et sp. indet., Kipchak Limestone (late Darriwilian – early Sandbian), Locality 166, NMW 98.28G.2053, dorsal exterior. **L.** *Altynorthis tabylgatensis* (Misius, 1986), Baigara Formation (late Darriwilian – early Sandbian), Locality 1026, NMW 98.28G.1973, ventral interior. Scale bars 2 mm.

Family Hesperomenidae Cooper, 1956

Genus *Anoptambonites* Williams, 1962

Type species (by original designation). – *Leptaena grayae* Davidson, 1883, from the Craighead Limestone (Sandbian), Girvan, Scotland.

Remarks. – *Anoptambonites grayae*, the type species of latest Sandbian age, is the only one named from the British Isles, although *Anoptambonites* sp. has also been described from several horizons in Scotland, then part of Laurentia (Cocks 2008). There are no records of the genus from the rest of Laurentia, Avalonia, or Baltica, but from Kazakhstan *Anoptambonites aktasensis* Nikitin & Popov *in* Klenina *et al.*, 1984, was named from the Bestamak Formation (Sandbian), *A. convexus* Popov, Cocks & Nikitin (2002) and *A. orientalis* Popov (1980b) were named from the Anderken Formation (Sandbian), *A. kovalevskii* Popov, Nikitin & Cocks, 2000, from the Dulankara Formation (early Katian), *A. perforatus*

Nikitin, Popov & Bassett, 2006, from the Angrensor Formation (early Katian), and *A. subcarinata* Nikitin & Popov, 1996, from the Sortan-Manai Mudmound (Katian).

Anoptambonites kovalevskii Popov, Nikitin & Cocks, 2000

Plate 11, figure 12; Figure 22F–H, K–R)

2000 *Anoptambonites kovalevskii* Popov, Nikitin & Cocks, p. 853, pl. 4, figs 9–18.

2006 *Anoptambonites kovalevskii* Popov, Nikitin & Cocks; Popov & Cocks, p. 269, pl. 8, fig. 27.

Holotype. – CNIGR 49/12375, complete shell, from Otar Member of the Dulankara Formation (early Katian) of Dulankara, southern Chu-Ili Range, Kazakhstan.

Material. – Kopkurgan Formation, Locality 735: six ventral and six dorsal valves mainly preserved as internal and external moulds, including BC 63812 (Figs 22M, N), BC 63819b, BC 63820a (Fig. 22K), BC 63835b (Fig. 22G), BC 63837a (Fig. 22R), BC 63838a (Fig. 22F), BC 63838d (Fig. 22O), BC 63841 (Fig. 22P), BC 63842 (Fig. 22L), BC 63878 (Fig. 22 H), average L, 15.9 (range 13.3-18.6), average W, 25.5 mm (range 21.8-28.4); dorsal moulds, BC 63886 (Pl. 11, fig. 12), BC 63888a, BC 63891a (Fig. 22Q), average L, 11.1 mm (range 6.0–14.8), average W 17.1 mm (range 9.2–23.8). Average ventral valve L/W ratio 0.63; dorsal valve L/W ratio 0.55.

Remarks. – Dimensions of the holotype CNIGR 49/12375 given when erecting *Anoptambonites kovalevskii*, were given as L, 13.4 mm and W, 17.5 mm, indicating a L/W ratio of 0.77, which is outside the range of the Baigara material noted here. However, as can be seen from its photograph (Popov *et al.* 2000, pl. 4, figs 11–13), the holotype has both alae missing, and thus its estimated width seems much more probably to have originally been about 22 mm, thus revising the L/W ratio as 0.61. That revision makes the identification of the new Baigara material as *Anoptambonites kovalevskii* much more secure.

Anoptambonites sp. 1

Figure 22J

Remarks. – A single small, subcarinate, ventral valve with parvicostellate radial ornament, BC 63843 from the Kopkurgan Formation at Locality 727, strongly recalls *Anoptambonites convexus* Popov *et al.*, 2002, which is common in the *Acculina-Dulankarella* and *Parastrophina-Kellerella* associations characteristic of the carbonate build-ups in the upper part of the Anderken Formation in the south Chu-Ili Range. However, the available material is insufficient for erecting a new species.

Anoptambonites sp. 2

Figure 22S–U

Material. – Locality 8126: dorsal valve BC 63808 (Fig. 22T); ventral valves BC 63809 (Fig. 22S), BC 63810 (Fig. 22U); BC 63811.

Remarks. – These shells came from the unnamed unit which disconformably overlies the Kopkurgan

Formation in the uppermost part of the Ordovician succession and the shells of *Anoptambonites* are the only identifiable fossils available from that unit. They are preserved in a distinctive shell bed deposited in high energy conditions and show variable amounts of exfoliation. The specimens show some similarity to *Anoptambonites kovalevskii* in their shell size and characters of radial ornament, but their species affiliation is uncertain due to the small quantity and poor preservation of the shells. They also have a younger, probably late Katian, age.

Genus *Chaganella* Nikitin, 1974

Type species (by original designation). – *Chaganella chaganensis* Nikitin, 1974, from the Bestamak Formation (lower Sandbian), Chagan River, Chingiz Mountains, Chingiz –Tarbagatai Terrane, Kazakhstan.

Chaganella chaganensis Nikitin, 1974

Figure 22A–E, I

1974 *Chaganella chaganensis* sp. nov. Nikitin, p. 343, pl. 6, figs 1–11.

Material. – Berkutsyur Formation, Locality 388: six ventral and eight dorsal valves, including BC 63790, BC 63791 (L, 11.3; W, 16.3; Fig. 22B), BC 63793 (L, 13.3; West, 21; Fig. 22D), BC 63794 (L, 11.8; W, 20.1; Fig. 22C), BC 63796 (Fig. 22A), BC 63800-806, BC 63807 (L, 13.1; W, 25.1; Fig. 22E, I), BC 63797-98.

Remarks. – The genus has only been found at a single late Sandbian locality in West Balkhash, and is its first record from the Chu-Ili Terrane. The specimens may be somewhat loosely referred to the type species, although only single examples of both ventral and dorsal valve interiors are known; the others only preserve the exteriors and are not well preserved.

Family Sowerbyellidae Öpik, 1930
Subfamily Sowerbyellinae Öpik, 1930

Genus *Sowerbyella* (*Sowerbyella*) Jones, 1928

Type species (by original designation). – *Leptaena sericea* de Sowerby 1839, from the Alternata Limestone Formation (Sandbian) of Shropshire, England.

Remarks. – The other subgenus of *Sowerbyella, S. (Rugosowerbyella)*, differs in the possession of small rugellae over the exterior shell surface and is not known from the Kazakh terranes, in contrast with *S. (Sowerbyella)*, which only has parvicostellate ornament, apart from occasional small rugae confined to the lateral extremities. A very large number of *Sowerbyella* species is known from around the world from rocks of Dapingian to Hirnantian ages, and their differentiation has been achieved in different ways by various authors; although the genus does not appear to have survived the Ordovician-Silurian boundary extinction events. In addition to *Sowerbyella (S.) ampla* and *S. (S.) verecunda* reviewed below, other Kazakh species are *Sowerbyella (S.) acculica* Misius, 1986, from the North Tien Shan Terrane, *S. (S.) akdombakensis* Klenina *in* Klenina *et al.*, 1984 [= *S. (S.) tamdysuensis* Misius, 1986], from the Chingiz-Tarbagatai, North Tien Shan, and Chu-Ili terranes, *S. (S.) insueta* sp. Klenina *in* Klenina *et al.*, 1984, *S. (S.) intricata* Nikiforova, 1978, *S. (S.) nativa*, Klenina *in* Klenina *et al.*, 1984, *S. (S.) plana* Klenina *in* Klenina *et al.*, 1984, *S. (S.) papiliuncula* Borissiak, 1972, *S. (S.) praestans* Klenina *in* Klenina *et al.*, 1984, all from the Chingiz-Tarbagatai Terrane; *S. (S.) rukavishnikovae* Popov, 1980, from the Chu-Ili Terrane; and *S. (S.) sinensis* Wang, *in* Wang & Jin (1964), which Nikitin *et al.* (2003) recorded from the Tauken Formation in the Selety Terrane.

Sowerbyella (Sowerbyella) ampla (Nikitin & Popov, 1996)

Plate 10, figures 17, 18; Plate 12, figures 7–10, 12–16

1996 *Anisopleurella ampla* sp. nov. Nikitin & Popov, p. 12, figs 5K–R.

2006 *Sowerbyella (Sowerbyella) ampla* (Nikitin & Popov); Popov & Cocks, p. 48, figs 12.7–12.14.

Holotype. – CNIGR 21/12877, a ventral valve from the Dulankara Mud Mound (Katian), Sartan Manal, northern Betpak Dala desert, Kazakhstan.

Alakol and North Betpak-Dala Material. – Kopkurgan Formation, Locality 735: three articulated shells 14 ventral and 18 dorsal valves, including BC 63824a (Pl. 12, fig. 12), BC 63838e (Pl. 12, figs 8, 9), BC 63839a (Pl. 12, fig. 16), BC 63840b (Pl. 12, fig. 7), BC 63844a (Pl. 12, figs 13, 14), BC 63844b (Pl. 12, fig. 15), BC 63882a (Pl. 12, fig. 10), BC 63889a. Unnamed late Ordovician Formation, Locality 1501:

two dorsal valves, NMW 98.28G.2065 (Pl. 10, fig. 17) and NMW 98.28G.2066 (Pl. 10, fig. 18).

Remarks. – *Sowerbyella (S.) ampla* is one of the most common species in the Late Ordovician of the Chu-Ili Terrane; and, in addition to its revision from the Dulankara Formation by Popov & Cocks (2006), Popov *et al.* (2002) recorded *S. (S.) aff. S. (S.) ampla* from the Anderken Formation, and Nikitin *et al.* (2006) the same from the Angrensor Formation in the Boshkekul Terrane. From West Balkhash, the species has only been identified from the middle Katian at Locality 735, while at North Betpak-Dala it was not previously reported from siliciclastic deposits.

Sowerbyella (Sowerbyella) verecunda verecunda Nikitin & Popov, 1983

(Not illustrated)

1983 *Sowerbyella verecunda* sp. nov. Nikitin & Popov, p. 239, pl. 3, figs 7–8, 10, 12, 16–17.

Holotype. – CNIGR 40/11960, ventral internal mould, from the Kreshchenovka Formation (early Sandbian) of Ishim River near Kupriyanovka Village, Kalmykkol-Kokchetav Terrane, Kazakhstan.

Sowerbyella (Sowerbyella) verecunda baigarensis n. subsp.

Plate 13, figures 2–17; Plate 15, figure 14

Derivation of name. – After Baigara Mountain near the type locality.

Holotype. – NMW 98.28G.1214 (Ld, 6.0; W, 10.2; Pl. 15, fig. 14), dorsal internal mould, from the Baigara Formation (late Darriwilian – early Sandbian), from Locality 1026, west side of Karatal River, southern Betpak-Dala; Kazakhstan.

Paratypes. – Baigara Formation, Locality 1026: six ventral and 12 dorsal valves, including NMW 98.28G.1215 (Pl. 13, fig. 8), 1215, BC 60862 (Pl. 13, fig 3).

Other material. – Berkutsyur Formation, Locality 812: two articulated shells, ten ventral and 6 dorsal valves, including BC 62325 (Pl. 13, fig. 14), BC 62326

(Pl. 13, fig. 13), BC 62327 (L, 5.1; W, 10.8), BC 62328 (L, 6.4; W, 12.2), BC 62337 (Pl. 13, fig. 7), BC 62338-40 (Pl. 13, fig. 4), NMW 98.28G.1217 (Pl. 13, fig. 16). Locality 813: four ventral valves, including BC 60819–22. Locality 815: two ventral and one dorsal valves, including NMW 98.28G.1219, 1220 (Pl. 13, fig. 12), 1999 (Pl. 13, fig. 11). Locality 816a: four ventral and four dorsal valves, including BC 60809–10, BC 62371–75. Locality 817: two articulated shells, 14 ventral and four dorsal valves, including BC 60789–90, BC 62395–96. Baigara Formation, Locality 1020a: One articulated shell, two dorsal and two ventral valves. Locality 1021: 51 articulated shells, six ventral and four dorsal valves, including NMW 98.28G.1151 (Pl. 13, fig. 17), 1170–1213, 1221, 1222 (Pl. 13, fig. 1), 1223 (Pl. 13, fig. 10), 1224 (Pl. 13, fig. 15), 1265–1273, 2123–2127. Locality 1022: 38 articulated shells. Locality 1023: 25 articulated shells. Locality 1025: four articulated shells. Locality 1026: BC 60863. Locality 1026b: two articulated shells. Locality 1028: four ventral and two dorsal valves, including NMW 98.28G.1150 (Pl. 13, fig. 2). Takyrsu Formation, Locality 166, four ventral and one dorsal valves, including NMW 98.28G.1152 (Pl. 13, fig. 5), 1153 (Pl. 13, fig. 6), 1154, 1155.1-4. Average sizes for 15 articulated shells from Locality1023. Length: X, 12.3; S, 2.68; min, 8.0; max, 17.0. Width: X, 20.9; S, 2.79; min, 17.0; max, 26.3. Thickness: X, 3.7; S, 0.95; min, 2.5; max, 5.2. L/W: X 61.1%, S, 15.5, OR, 38.5-82.4%. T/L: X, 32.0%, S, 12.8; OR, 19.4–65.0%).

Diagnosis. – Concavoconvex shell about three-fifths as long as wide. Radial ornament with 9–11 accentuated ribs originating at the umbo and 5–8 parvicostellae per mm along the anterior margin. Dorsal central side septa prominent, subparallel, terminating at the anterior margin of a weakly defined bema together with a long, faint median septum. Dorsal outer side septa variably developed.

Description. – Shell concavoconvex, transverse, semioval outline, about three-fifths as long as wide with maximum width at the hinge line, and about one-third as thick as long. Cardinal extremities acute to almost right angled. Ventral valve moderately and evenly convex, with a low, apsacline interarea, bearing a small, apical pseudodeltidium. Dorsal valve gently concave with the low planar hypercline interarea. Notothyrium covered by the chilidial plates merged at the umbo. Radial ornament inequally parvicostellate with 9–11 accentuated ribs originating at the umbo and with two generations of secondary ribs inclined at about mid-length and close to the shell margins of the largest individuals (Pl. 13, figs 12, 14). Interspaces between parvicostellae contain 4–6 parvicostellae at mid-length and 2–4 along the anterior margin of the mature individuals with 5–8 parvicostellae along the anterior margin in 78% of the specimens. Faint, very weakly defined oblique rugellae present along the hinge line of some individuals, nevertheless those rugellae are not characteristic. Ventral valve interior with small transverse teeth and short widely divergent dental plates. Muscle field bilobed, open anteriorly, bounded laterally by prominent muscle boulding ridges. Diductor muscle scars large, extending anteriorly almost to the mid-valve, adductor scars significantly shorter, lanceolate, divided by a median ridge, enclosed completely by the diductor scars. Ventral mantle canals strongly impressed, lemniscate. Dorsal interior with undercut cardinal process, widely divergent socket plates bounding anteriorly deep, transverse sockets (Pl. 13, fig. 13). Bema weakly defined, undivided. Central side septa faint posteriorly becoming prominent anteriorly, narrowly divergent to almost subparallel extending anteriorly towards the margin of the bema and usually terminating at about three-quarters valve length. Median septum faint and long, extending anteriorly to the margin of the bema. Outer side septa variably developed, usually faint and short (Pl. 13, figs 7, 13; Pl. 15, fig. 14). Mantle canal system lemniscate (Pl. 13, fig. 16).

Remarks. – *Sowerbyella* populations from the West Balkhash Region differ consistently from those with slightly less transverse shells (L/W on average 59–61% against 55%) and slightly coarser radial ornament (5–8 per mm against 7–11) and a less prominent bema. While the listed differences of *Sowerbyella* (*S.*) *verecunda* from the West Balkhash cannot be considered significant enough to distinguish them from typical *Sowerbyella* (*S.*) *verecunda* on a species level, nevertheless and keeping in mind their considerable geographical isolation, the West Balkhash and Kokchetav-Kalmykkol populations are differentiated here at subspecies level. All of the above localities from the West Balkhash region are of early Sandbian ages, apart from localities 1021–1023, which are late Darriwilian. However, in addition there are a few specimens (e.g. from Locality 538), which, although only identifiable as *Sowerbyella* (*Sowerbyella*) sp., are of late Sandbian age.

Sowerbyella (*Sowerbyella*) cf. *S.* (*S.*) *acculica* Misius, 1986

Plate 15, figures 12, 13

cf. 1986 *Sowerbyella* (*Sowerbyella*) *acculica* sp. nov. Misius, p. 154, pl. 14, fig. 14–32.

Material. – Berkutsyur Formation, Locality 816: ventral valve BC 62373 (Pl. 15, figs 12, 13). Locality 8121: one ventral valve.

Remarks. – There are a few *Sowerbyella* shells which can be easily distinguished from *Sowerbyella (S.) verecunda* in having a broad, flat-bottomed ventral median sulcus and a top-flat dorsal median fold bounded by plications and numerous straight oblique rugellae along the hinge line, and clearly represent a different taxon. Similar shells which are also characterised by a broad, flat ventral sulcus bordered by a pair of lateral plications were described by Misius (1986, p. 154) under the name *Sowerbyella acculica* from the middle part of the Tabylgaty Formation (Sandbian) of the Moldo-Too Range in North Kyrgyzstan (North Tien Shan Terrane). The description in that publication suggests that the Chu-Ili Terrane shells are probably conspecific, but because of poor original illustrations and lack of access to topotypes, it cannot be proved.

Genus *Gunningblandella* Percival, 1979

Type species (by original designation). – *Gunningblandella resupinata* Percival, 1979, from the Goonumbla Volcanics (Katian), New South Wales, Australia.

Gunningblandella sp. 1

Plate 12, figures 2–6, 11

2006 *Gunningblandella* sp. Popov & Cocks, p. 271, pl. 5, figs 22, 23, 26.

Material. – Dulankara Formation, Degeres Member, Locality 828: external mould of conjoined valves, BC 57710 (L, 8.6, W, 14.5); Locality 832: ventral internal mould, BC 57712 (L = 8.2, W = 13.1). Kopkurgan Formation, Locality 735: one articulated shell, BC 63837c (Pl. 12, fig. 5); 10 ventral and six dorsal valve preserved as internal and external moulds, including, BC 63827b, BC 63829 (Pl. 12, fig. 3), BC 63830c (Pl. 12, fig. 2), BC 63837d (Pl. 13, fig. 11), BC 63838b, BC 63840a (Pl. 12, fig. 4), BC 63874a (Lv, 8.9; W, 17.3; Pl. 12, fig. 6), BC 63874c.

Remarks. – Ordovician sowerbyellids were reviewed by Cocks (2013), and *Gunningblandella* stands out as being the only member of the family which is resupinate. However, it is relatively rare, being only known so far from its type locality in Australia (where it is abundant), from two specimens identified as *Gunningblandella* sp. from Localities 828 and 832 in the Dulankara Formation (Popov & Cocks 2006), and now from the new material here from Locality 735 in the Berkutsyur Formation, which are all within the Chu-Ili Terrane and of Katian ages. The specimens from Locality 735 are approximately contemporaneous and probably conspecific with those from the Dulankara Formation. They differ from *G. resupinata* in being rather less resupinate than the type species. Other differences include a broadly uniplicate anterior commissure, prominent ventral sulcus and dorsal median fold anterior to mid-length (Pl. 12, figs 5, 6), which suggest a separate species; however, in the absence of data on the dorsal interior we prefer to keep it in open nomenclature.

Gunningblandella sp. 2

Plate 15, figure 16

Remarks. – There is a single well-preserved ventral internal mould from Locality 538 in the Anderken Formation, BC 32727, which was not included in the Anderken fauna described by Popov *et al.* (2002), which is definitely *Gunningblandella*, but it is of late Sandbian age, making it the oldest-known specimen of the genus yet recorded. The specimen has a dorsal median sulcus recalling the shells from the Kopkurgan Formation discussed above, but the available material is not enough to discuss its taxonomic affinity in more detail.

Genus *Ptychoglyptus* Willard, 1928

Type species (by original designation). – *Ptychoglyptus virginiensis* Willard, 1928, from the Edinburg Formation (Sandbian) of Virginia, USA.

Ptychoglyptus sp.

Plate 7, figures 19, 23

Material. – Berkutsyur Formation, Locality 8121: ventral valve, BC 62403 (Pl. 7, fig. 23). Takyrsu Formation, Locality 166: ventral valve, BC 62443 (Pl. 7, fig. 19), ventral external mould, BC 62444.

Remarks. – This planoconvex shell with characteristic surface ornament of three accentuated ribs radiating from the umbo with faint parvicostellae in the interspaces and numerous concentric rugae interrupted by accentuated ribs leave no doubts about the generic affiliation of these shells (see also Cocks & Rong 1989). However, no interiors are known and the inadequate material makes species attribution impossible. The only other record of *Ptychoglyptus* in Kazakhstan is from the Chingiz Terrane (Nikitin & Popov, 1984), where an unnamed species was reported from the upper Bestamak Formation (Sandbian).

Order Orthotetida Waagen, 1884
Suborder Orthotetidina Waagen, 1884
Superfamily Chilidiopsoidea Boucot, 1959
Family Chilidiopsidae Boucot, 1959
Subfamily Chilidiopsinae Boucot, 1959

Genus *Gacella* Williams, 1962

Type species (by original designation). – *Gacella insolita* Williams, 1962, from the Stinchar Limestone Formation (Sandbian) of Girvan, Scotland.

Gacella sp.

Plate 12, figure 1

Material. – Kopkurgan Formation, Locality 735: BC 63822c (Lv, 8.3; W, 9.6; Ml, 3.2), ventral internal mould.

Remarks. – A single ventral internal mould shows a strongly elongate muscle field bisected anteriorly by a low median ridge and bounded laterally by long, subparallel dental plates. No exteriors and dorsal valves are known, nevertheless the ventral interior indicates attribution to *Gacella*. This is the earliest documented occurrence of the genus in Kazakhstan, since all the previous records of Kazakh representatives of the genus are confined to the Sandbian Stage. In particular, *Gacella institata* Popov *et al.*, 2002 occurs in the Anderken Formation of the Chu-Ili Terrane; also *Gacella sulcata* Misius (*in* Misius & Ushatinskaya 1977) is known from the Tabylgaty Formation in the North Tien Shan Terrane.

Suborder Triplesiidina Moore, 1952
Superfamily Triplesioidea Schuchert, 1913
Family Triplesiidae Schuchert, 1913

Genus *Grammoplecia* Wright & Jaanusson, 1993

Type species (by original designation). – *Grammoplecia triplesioides* Wright & Jaanusson, 1993, from the Boda Limestone (late Katian) of Dalarna, Sweden.

Grammoplecia globosa (Nikitin & Popov, 1985)

Plate 14, figures 3, 6–17; Figure 23A–D

1985 *Triplesia globosa* sp. nov. Nikitin & Popov, p. 40, pl. 2, figs 1–6.

Holotype. – CNIGR 23/12209, a pair of conjoined valves from the Andryushinskaya Formation (Sandbian) of north-western Central Kazakhstan.

Material. – Baigara Formation, Locality 1026: NMW 98.28G.1459-65; one articulated shell, 26 ventral and 24 dorsal valves. Locality 1026b: 67 articulated shells, including NMW 98.28G.1397, (Pl. 14, fig. 3), 1398 (Pl. 14, figs 7, 10, 12), 1399 (Pl. 14, figs 6, 8, 9, 11), 1400 (Pl. 14, figs 14-17), 1401–1457, 1462 (Figs 23A-D). Average sizes for articulated shells from Locality1026b. Length: X, 13.9; S, 1.88; min, 9.1; max, 18.9; N, 62. Width: X, 15.9; S, 2.12; min, 10.8; max, 20.5, N, 62. Thickness: X, 8.7; S, 1.48; min, 5.5; max, 11.9, N, 56. Sw: X, 10.4; S, 1.51; min, 5.5; max, 14.8, N, 56. L/W: X 63.6%, S, 11.2, OR, 44.1–93.6%; N, 62. T/L X, 65.5%, S, 7.2; OR, 47.4–83.6%, N, 56). St/Sw: X, 65.5%; S, 7.2; min. 47.4%, max, 83.6%.

Remarks. – In size, proportions (apart from a somewhat lower thickness to length ratio) and finely multicostellate radial ornament with 4–5 ribs per mm, the shells from the Baigara Formation of the Chu-Ili Terrane and the Andryushinskaya Formation of the Kokchetav-Kalmykkol Terrane are almost indistinguishable (Nikitin & Popov 1985), and are very close in average shell sizes and proportions although the shells from the Baigara Formations are on average slightly less elongate (87.8% against 84%). The interior was originally described from a transverse serial section, but ventral internal moulds in the newly available material (Pl. 14, fig. 3) show a subtriangular muscle field open anteriorly and bounded laterally by thin, straight, divergent dental plates. No individual muscle scars are preserved. *Grammoplecia globosa* also has a pseudodeltidium with a monticulus and a keeled cardinal process with cowl; these characteristic features were not included in the original description.

Grammoplecia aff. *G. globosa* (Nikitin & Popov, 1985)

Plate 14, figures 1–2, 4, 5

Material. – Baigara Formation, Locality 1022: nine articulated shells. Locality 1023: nine articulated shells, including NMW 98.28G.1381, 1382 (Pl. 14, figs 1, 2, 4, 5), 1389–1395. Locality 625-1: seven articulated shells, including NMW 98.28G.1383–1388. Average sizes for articulated shells from the Baigara Formation. Length: X, 13.2; S, 2.60; min, 9.1; max, 17.6. Width: X, 16.6; S, 3.17; min, 10.4; max, 24.4. Thickness: X, 5.9; S, 1.0; min, 4.5; max, 13.0; N, 19. Sw: X, 9.4; S, 1.93; min, 5.4; max, 13.0; N 18. L/W: X, 84.6%; S, 17.5; min, 59.9%; max, 120.8%; N, 19. Th/L: X, 43.9%; S, 11.1; min, 28.7%; max, 80.4%; N, 18. Sw/W: X, 57.1%; S, 6.5; min, 45.5%; max, 70.3%, N, 18.

Remarks. – Specimens of *Grammoplecia* from the lower part of the Baigara Formation represent the earliest known triplesioid from Kazakhstan, While their finely multicostellate radial ornament (4–5 ribs per mm) are similar to *Grammoplecia globosa*, they show a statistically significant difference in average T/L ratio from typical *Grammoplecia globosa* (44% against 65% in the types and 65.5% in the population from Locality 1026b). They also differ from the latter taxon in lacking a swollen umbonal region in the dorsal valve, a less prominent dorsal median fold, and a dorsal sulcus terminating in a low, subtrapezoidal tongue. The specimens from the Baigara Formation are variably distorted, but it is likely that the obtained average values give a fair representation of the original shell size and proportions, and therefore the observed differences may have taxonomical significance; however, due to the imperfect preservation of most of the shells, it seems best to leave them in open nomenclature.

Grammoplecia sp.

Figure 24E, F, K

Material. – Kopkurgan Formation, Locality 735: six ventral internal moulds, BC 63819a, BC 63824e, BC 63826 (Fig. 24F), BC 63873c, BC 63885b (Fig. 24K); one incomplete ventral external mould, BC 63819a (Fig. 24E).

Remarks. – The only previously documented species of the genus from the Katian of the Chu-Ili Terrane is *Grammoplecia subcraegensis* (Rukavishnikova, 1956), which occurs occasionally in the Dulankara Formation. The specimens from the Kopkurgan Formation recall the former taxon in their large sizes, finely capillate radial ornament, and well defined ventral sulcus with steep lateral slopes. The major differences are in the ventral valve interior. *Grammoplecia subcraegensis* has relatively short, thin, widely divergent dental plates and a slightly elongate suboval, gently impressed ventral muscle field bisected by a long median ridge (e.g. Popov *et al.* 2000, pl. 1, fig. 10), while in the specimens from the Kopkurgan Formation the dental plates are longer and less divergent, the ventral muscle field is not clearly impressed, and the median ridge is absent. In the dental plates and muscle field they are more comparable to *Grammoplecia wrighti* Popov *et al.*, 2002, from the Anderken Formation (late Sandbian) of Chu-Ili, but differ in the less strongly laterally defined ventral sulcus originating at some distance from the beak. In the absence of data on the dorsal valve morphology, the species of the specimens remains uncertain.

Genus *Triplesia* Hall, 1859

Type species (by original designation). – *Atrypa extans* Emmons, 1842, from Sandbian age rocks in New York State, USA.

Triplesia sp.

Plate 14, figures 18–20

Remarks. – A single pair of conjoined smooth articulated shells, BC 63698 from the Berkutsyur Formation at Locality 8121, probably represent an immature individual of *Triplesia*. The specimen is characterised by a weakly uniplicate anterior commissure and a pseudodeltidium bisected by the monticulus, and a weak dorsal median fold and ventral sulcus originating close to the anterior margin. The specimen may be conspecific with the shells described by Popov *et al.* (2002) as *Triplesia* aff. *T. subcarinata* Cooper, 1956, but the material is insufficient to be sure.

Order Billingsellida Schuchert, 1893

Superfamily Polytoechioidea Opik, 1934

Family Tritoechiidae Ulrich & Cooper, 1936

Genus *Eremotoechia* Cooper, 1956

Type species (by original designation). – *Eremotoechia cloudi* Cooper, 1956, from the Arline Formation (Sandbian) of Tennessee, USA.

***Eremotoechia inchoata* Popov, Vinn & Nikitina, 2001**

(Not figured)

2001 *Eremotoechia inchoata* nov. sp. Popov, Vinn & Nikitina, p. 143, figs 8.8–21.

Holotype. – NMW 98.28G.35, articulated shell, from Locality 1023, Baigara Formation (late Darriwilian to early Sandbian), about 6 km SW of Baigara Mountain, south Betpak-Dala.

Remarks. – Detailed description and discussion of this species was published by Popov *et al.* (2001). *Eremotoechia* is the only polytoechoid genus yet documented from the Darriwilian and Sandbian of the Kazakh terranes and the species is endemic to the Chu-Ili Terrane.

***Eremotoechia spissa* Popov, Vinn & Nikitina, 2001**

(Not figured)

2001 *Eremotoechia spissa* – nov. sp. Popov, Vinn & Nikitina, p. 143, figs 8.1–7, 9.5–7.

Holotype. – By original designation, NMW 98.28G. 30, articulated shell, from Locality 1026b, Baigara Formation (late Darriwilian – early Sandbian), west side of Karatal River, southern Betpak-Dala, Kazakhstan.

Remarks. – Detailed description and discussion of the species were published by Popov *et al.* (2001).

Order Protorthida Schuchert & Cooper, 1931

Suborder Clitambonitidina Öpik, 1934

Remarks. – As noted by Popov *et al.* (2001), the Clitambonitidina (excluding the Polytoechioidea) are not closely related to the Billingsellida and should be considered as derived protorthides or less probably as a sister group of the orthides, a conclusion which was confirmed by comparative ontogenetic studies of polytoechioids and clitambonitoids. Therefore, the Suborder Clitambonitidina (*sensu stricto*) is assigned here to the Order Protorthida.

Superfamily Clitambonitoidea Winchell & Schuchert, 1893

Family Clitambonitidae Winchell & Schuchert, 1893

Genus *Atelelasma* Cooper, 1956

Type species (by original designation). – *Atelelasma perfectum* Cooper, 1956, from the Arline Formation (Sandbian) of Tennessee, USA.

***Atelelasma* sp.**

Figure 24A–D, G, H

Material. – Baigara Formation, Locality 1026b: two conjoined valves, NMW 98.28G.2037 (Fig. 24A–C, G) and 2038 (Fig. 24D, H).

Description. – These specimens are characterised by a planoconvex, transverse, subrectangular shell with subpyramidal ventral valve and strongly flattened dorsal valve ornamented by fine costellae, which are poorly preserved due to strong exfoliation of the shells. The ventral interarea is high, procline with a broad, triangular, open delthyrium, while the ventral interior is characterised by the presence of a spondylium. All these features suggest attribution to *Atelelasma*. In having a relatively small shell with an almost flat dorsal valve and a procline ventral interarea of a subpyramidal ventral valve they recall *Atelelasma planum* Cooper, 1956, from the Sandbian Benbolt Formation of Tennessee, but differ in their strongly transverse shell and in the lack of a dorsal sulcus. Another similar species is *Atelelasma peregrinum* Andreeva *in* Nikiforova & Andreeva, 1961, from the Volginsk Regional Stage of Central Siberia, but the latter is significantly larger and has a distinctly convex dorsal valve. Although there is no other record of clitambonitoids in the Ordovician of Kazakhstan, the specimens are left in open nomenclature due to poor preservation and lack of data on the dorsal interior.

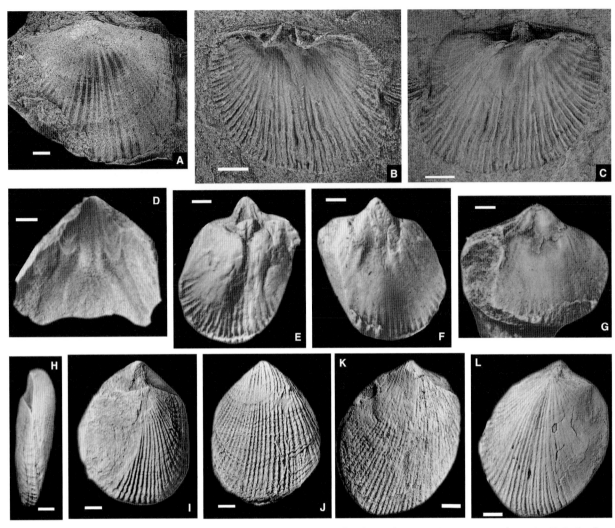

Fig. 25. **A.** *Altynorthis vinogradovae* n. sp. Berkutsyur Formation (early Sandbian), Locality 813; BC 60827, ventral exterior. **B–L.** *Scaphorthis recurva* Nikitina, 1985; **B–C**, near Talapty winter hut, Rgaity river basin, South Kendyktas Range, Tien Shan, Locality 12211; NMW 98.66G.2075, dorsal interior, internal mould; **D–L**, Baigara Formation (late Darriwilian – early Sandbian), Locality 1021; D, NMW 98.28G.1869, ventral interior; **E, F**, NMW 98.28G.1874, internal mould of conjoined valves,; **G**, NMW 98.28G.1873, ventral internal mould; **H–J**, NMW 98.28G.1870, exterior views of conjoined valves; **K, L**, NMW 98.28G.1871, exterior views of conjoined valves. Scale bars 2 mm.

Order Orthida Schuchert & Cooper, 1932

Suborder Orthidina Schuchert & Cooper, 1932

Superfamily Orthoidea Woodward, 1852

Family Hesperorthidae Schuchert & Cooper, 1931

Subfamily Dolerorthinae Öpik, 1934

Genus *Dolerorthis* Schuchert & Cooper, 1931

Type species (by original designation). – *Orthis interplicata* Foerste, 1909, from the Osgood Formation (Llandovery: Telychian) of Indiana, USA.

***Dolerorthis expressa* Popov, 1980b**

Plate 15, figures 6, 7

1980b *Dolerorthis expressa* sp. nov. Popov, p. 144, pl. 1, figs 5–7.

2002 *Dolerorthis expressa* Popov; Popov, Cocks & Nikitin, p. 58, pl. 1, fig. 29, pl. 11, figs 1, 2.

Holotype. – CNIGR 11/11523, ventral internal and external moulds, from Locality 1018, the Anderken Formation (Sandbian), from about 7 km south-west of Karpkuduk well, Kotnak Mountains, southern Bet-pak-Dala, Kazakhstan.

Remarks. – A single exceptionally well preserved dorsal internal mould, BC 62387 from the Berkut-syur Formation at Locality 817, shows strongly impressed apocopate mantle canals characteristic of the type species, but not previously reported for *Dolerorthis expressa*. The affinity of the species was discussed by Popov *et al.* (2002).

Genus *Sonculina* Misius, 1986

Type species (by original designation). – *Sonculina prima* Misius, 1986, from the Tabylgaty Formation (Sandbian, *Leptellina* Beds) of the Moldo-Too Range, north Kyrgyzstan (North Tien Shan Terrane).

Diagnosis. – Shell subequally biconvex, with hinge line shorter than maximum shell width at mid-length and gently uniplicate to rectimarginate anterior commissure. Ventral interarea gently curved, apsacline with an open delthyrium, shallow ventral sulcus present close to the anterior margin. Dorsal valve with a low, planar, anacline interarea, shallow sulcus originating at the umbo and inverted into a low median fold close to the anterior margin. Radial ornament multicostellate. Ventral interior with short divergent dental plates, slightly elongate suboval muscle field, with anteriorly raised adductor scars longer than diductor scars; ventral mantle canals saccate with prominent, subparallel *vascula media*. Dorsal valve interior with a simple ridge-like cardinal process on the low notothyrial platform, high, divergent, blade-like brachiophores, and weakly impressed dorsal adductor muscle scars divided by a strong and broad dorsal median ridge. Dorsal mantle canals apocopate.

Remarks. – Williams & Harper (2000) put this genus into the synonymy of *Mimella* Cooper, 1930; however, our study suggests that they are not congeneric, and *Sonculina*, because of the morphology of the ventral muscle field and cardinalia as revised here, represents a member of the Hesperorthidae. Among the early genera of the family it recalls *Paradolerorthis* Zeng, 1987, in having a subquadrate biconvex shell with obtuse cardinal extremities, a similar ventral muscle field and prominent subparallel *vascula media*, but it differs in having a variably uniplicate anterior commissure and finer multicostellate radial ornament. *Sonculina* is also very similar to early species of *Dolerorthis*, and their differences are relatively minor, including a subequally biconvex shell, suboval ventral muscle field with longer adductor scar, finer multicostellate radial ornament and

more distinctly uniplicate anterior commissure. *Sonculina* also has a relatively well-defined ventral sulcus and dorsal median fold.

Sonculina cf. *S. prima* Misius, 1986

Plate 16, figures 2–7, 9

cf. 1986 *Sonculina prima* gen. et sp. nov. Misius, p. 113, pl. 8, figs 1–20, pl. 9, figs 1–17.

Material. – Baigara Formation, Locality 1026b, 16 articulated shells, including NMW 98.28G.1147 (Pl. 16, figs 2–4), 1148 (Pl. 16, figs 5, 9), 1149 (Pl. 16, figs 6, 7). Average sizes for articulated shells from Locality1026b. Length: X, 9.9; S, 1.56; min, 6.7; max, 12.1; N, 9. Width: X, 11.6; S, 2.92; min, 7.3; max, 17.3; N, 8. Thickness: X, 5.1; S, 0.93; min, 3.0; max, 6.2; N, 9.

Description. – Shell strongly and slightly ventribiconvex, slightly transverse, subrectangular, on average 88% (S, 10.7; OR, 70–99%; N, 9) as long as wide with maximum width at a mid-length and 52% (S, 9.4; OR, 40–67%) as thick as long. Hinge line considerably shorter than maximum shell width, cardinal extremities obtuse. Anterior commissure gently uniplicate. Ventral valve with an erect beak posteriorly and slightly curved posterodorsally; lateral profile and strongly evenly convex with the maximum height slightly posterior to mid-length. Ventral interarea high, apsacline, gently curved in cross-section, with open triangular delthyrium. A weakly-defined sulcus originates anterior to mid-length. Dorsal valve with slightly swollen umbonal area and moderately convex lateral profile with maximum height at about quarter valve length from the umbo. Shallow sulcus between the umbo and mid-valve, becoming inverted into a low and broad median fold anterior to mid-length. Radial ornament multicostellate with 8–10 per 3 mm ribs along the anterior margin of mature individuals. A few thin concentric growth lamellae may be present along the anterior and lateral margins of larger shells. Ventral interior with teeth supported by strong and short, anteriorly divergent dental plates. Ventral muscle field strongly impressed slightly elongate suboval with strongly raised anteriorly adductor scars, slightly longer than diductor scars. Ventral mantle canals with strongly impressed, subparallel *vascula media*. Dorsal interior with a simple, ridge-like cardinal process on a low notothyrial platform and high, divergent, blade-like brachiophores. Dorsal adductor muscle scars weakly impressed and divided by a strong and broad dorsal median ridge.

Remarks. – The shells from the Baigara Formation are very similar in their exteriors to *Sonculina prima*, including sagittal profile, interareas, the anterior position of dorsal median fold and ventral sulcus, and also their radial ornament. The provisional species assignment of the Kazakh shells is mainly because of inadequate preservation of the dorsal interior.

Sonculina baigarensis n. sp.

Plate 16, figures 8, 10–16, 18–21; Plate 20, figures 1, 2

Derivation of name. – After Baigara Mountain, near the type locality.

Holotype. – NMW 98.28G.1889 (Lv, 19.9; Ld, 16.1; W, 23.5; T, 8.3; Pl. 16, figs 19–21), an articulated shell from Locality 625/3, Baigara Formation (late Darriwilian to early Sandbian), about 6 km SW of Baigara Mountain, south Betpak-Dala, South Kazakhstan.

Paratypes. – Locality 1021: six articulated shells and two dorsal valves, including NMW 98.28G.1846 (Ld, 11.8; W, 17.7; Iw, 12.1; Ml, 6.0; Mw, 8.5; Pl. 16, fig. 8). Locality 1022: 26 articulated shells, including NMW 98.28G.1888 (Pl. 16, figs 12-14), 1822–1845, 2121 (Pl. 16, figs 10, 11), 2122 (Lv, 14.8; Ld, 13.5; W, 16.0; Iw, 15.2; Pl. 20, figs 1, 2). Locality 625/3: eight articulated shells, including NMW 98.28G. 1890 (Lv, 16.7; Ld, 14.9; W, 18.3; T, 8.9; Iw, 12.4; Pl. 16, figs 15, 16, 18). Locality 1023: 12 articulated shells, including NMW 98.28G.1878–1887. Locality 1025: 2 articulated shells.

Diagnosis. – Subequally biconvex shell usually with gently uniplicate anterior commissure, dorsal median sulcus originating at the umbo, fading anterior to mid-length and inverted into a low median fold close to anterior margin. Radial ornament multicostellate with 25–30 ribs at the umbo and up to 60 ribs in larger individuals and 2–4 ribs per mm along the anterior margin.

Description. – Shell subequally biconvex, outline subrectangular to almost subcircular, on average 87% (S, 20, N, 23) as long as wide with maximum width at mid-length and about 43% (S, 11, N, 23) as thick as long; hinge line shorter than maximum shell width at mid-length; cardinal extremities obtuse. Anterior commissure gently uniplicate to rectimarginate. Ventral valve with small, pointed beak, curved

posterodorsally. Sagittal profile of ventral valve moderately and unevenly convex with maximum height at one-third valve length. Ventral interarea high, apsacline, gently curved, with the open triangular delthyrium and vestigial deltidial plates on its flanks (Pl. 16, fig. 20). Dorsal valve with a swollen umbonal area and a sagittal profile with maximum height at one-third valve length. Dorsal interarea strongly anacline. Shallow sulcus originating at the umbo, gradually fading towards the mid part of the shell and inverted into a broad, but weakly defined median fold between the mid-length and the anterior margin. Multicostellate radial ornament with 25–30 ribs at the umbo and 50-60 ribs along the anterior and lateral shell margins and 4–6 ribs per 3 mm along the anterior margin of larger individuals. Radial ornament with dense, faint, ridge-like concentric fila, about 3 or 4 per mm. Ventral valve interior with teeth supported by short, widely divergent dental plates extended anteriorly into muscle bounding ridges. Ventral muscle field slightly elongate, suboval, between one-third and two-fifths valve length, extending slightly beyond the delthyrial cavity floor, raised anteriorly. Ventral adductor muscle scar broad, triangular, about equal width, but slightly longer than gently impressed diductor scars. Ventral mantle canal system saccate with prominent, subparallel distally but curved outward anteriorly, impressions of *vascula media* and *vascula arcuata* diverging at just under right angles close to the anterior margin (Pl. 20, fig. 2). Dorsal interior with a simple ridge-like cardinal process on a low, subtriangular notothyrial platform and low, blade-like, widely divergent brachiophores (Pl. 16, fig. 8). Dorsal adductor scars gently impressed with a larger anterior pair, bisected by a low broad median ridge. Dorsal mantle canals apocopate.

Remarks. – This species differs from *Sonculina prima*, which is the type and only other known species of the genus, in having a less developed dorsal median fold and ventral sulcus in the anterior part of the valve and a more prominent dorsal sulcus posterior to the mid-length, as well as coarser radial ornament, with 3-6 ribs per 3 mm.

Hesperorthidae gen. et sp. indet.

Plate 15, figures 8, 11, 15; Figure 24J

Material. – Takyrsu Formation, Locality 166: three ventral and seven dorsal valves, including NMW 98.28G.2052 (Pl 15, fig. 8), 2053 (Fig. 24J), 2054 (Pl. 15, fig. 11), 2055 (Pl. 15, fig. 15), 2056–2059, 2062, 2063.

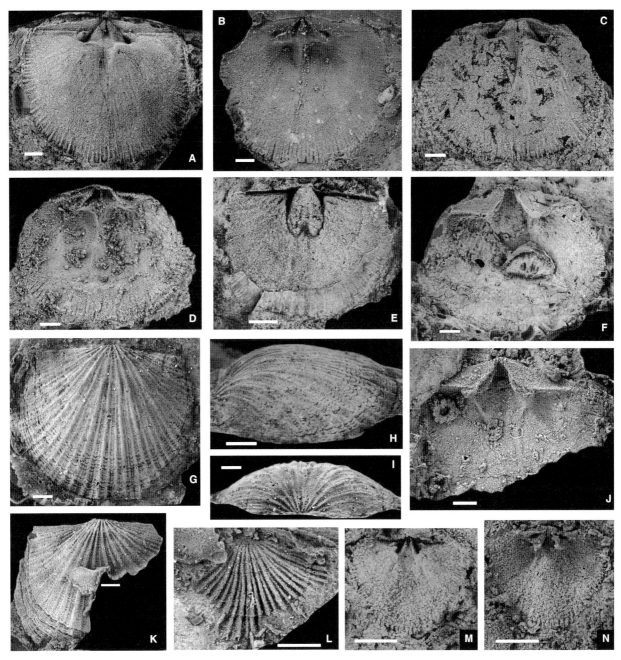

Fig. 26. **A–K.** *Phaceloorthis corrugata* n. sp.; Kopkurgan Formation (Katian), Locality 735; **A, B,** BC 63876b, holotype, dorsal internal mould and latex cast of dorsal interior; **C, D,** BC 63821a, dorsal internal mould and latex cast of dorsal interior; **E,** BC 63883, ventral internal mould; **F,** BC 63820d, latex cast of ventral interior; **G–I,** BC 63820c, latex cast of dorsal exterior, dorsal, lateral and posterior views; **J,** BC 63875, latex cast of ventral interior; **K,** BC 63839b, latex cast of dorsal exterior. **L–N.** *Onniella*? sp.; Kopkurgan Formation (Katian), Locality 735; **L,** BC 63839c, latex cast of incomplete ventral exterior; **M, N,** BC 63888c, dorsal internal mould and latex cast of exterior. Scale bars 2 mm.

Remarks. – These are slightly ventribiconvex, transverse, subrectangular to semioval shells with hinge line almost equal to maximum width at mid-length and weakly fascicostellate radial ornament (Fig. 23J). Ventral interior with slightly elongate, suboval, anteriorly raised muscle field occupying slightly more than one-quarter valve length, extending slightly

beyond the delthyrial cavity floor with adductor muscle scar slightly longer than diductor scars (Pl. 15, fig. 15). Ventral mantle canals saccate with *vascula media* subparallel in front of the muscle field, then widely divergent anteriorly (Pl. 15, fig. 2). In the ventral interior these shells recall *Sonculina*, but differ in their weakly fascicostellate radial ornament and a considerably smaller ventral muscle field. In the absence of data on the dorsal interior, generic discrimination cannot be made.

Family Glyptorthidae Schuchert & Cooper, 1931

Genus *Eridorthis* Foerste, 1909

Type species (by original designation). – Plectorthis (Eridorthis) nicklesi Foerste, 1909, from early Katian beds in Kentucky, USA.

Remarks. – According to Williams & Harper (2000), the only difference between the two very closely-related late Ordovician genera *Eridorthis* and the hitherto more widely-quoted *Glyptorthis* Foerste, 1914, whose type species is *Orthis insculpta* Hall, 1847, from early Katian beds in Ohio, is the anterior dorsal median fold in *Eridorthis* and the sulcus in *Glyptorthis*. However, as discussed more fully by Cocks (2019), in many species the fold and sulcus can be absent, slightly ventrally, or dorsally directed within individuals within the same population, and thus the two genera were placed in synonymy.

Eridorthis sp.

Plate 17, figures 1–12

Material. – Takyrsu Formation: Locality 166: one ventral and two dorsal valves, including NMW 98.28G.1997 (Pl. 17, fig. 9), 1998, 2036. Berkutsyur Formation: Locality 388: two ventral valves, including BC 63792 and BC 63795. Locality 8120/4b: one dorsal valve, BC 62399. Locality 8233: six articulated shells and one ventral valve, including NMW 98.28G. 1999, 2010 (Pl. 17, figs 1, 5), 2011 (Lv, 6.8; Ld, 6.6; W, 9.5; Iw, 7.7; T, 4.6; Pl 17, figs 2, 8, 10), 2012 (Pl 17, figs 3, 7, 11), 2013 (Lv, 6.6; Ld, 6.6; W, 8.8; T, 4.3; Pl 17, figs 4, 8, 12), 2014, 2015. Locality 8236a: two articulated shells, including BC 60812–13. Locality 8124: three articulated shells, BC 63702–4.

Description. – Shell ventribiconvex, transverse, rounded to subrectangular in outline, about three-quarters as long as wide and one-third as thick as long. Hinge line slightly shorter than maximum shell width at mid-length; cardinal extremities slightly obtuse. Anterior commissure slightly uniplicate to almost rectimarginate. Ventral valve subpyramidal

with sagittal profile almost straight to gently convex anterior to the umbo (Pl. 17, figs 10, 11, 12). Ventral interarea high, subtriangular, planar, procline to slightly apsacline with a narrow, open triangular delthyrium. Shallow ventral sulcus anterior to mid-valve. Dorsal valve profile moderately and evenly convex with maximum height slightly posterior to mid-length. Dorsal interarea low, planar, orthocline with narrow, open notothyrium. Dorsal umbo blind, slightly swollen, bisected by a shallow median sulcus fading to the mid-valve, where it is replaced by an indistinct median fold (Pl 17, figs 5, 6). Radial ornament ramicostellate, 20–24 high, subangular ribs increasing by bifurcation and separated by interspaces of equal width. Concentric ornament of strong, evenly spaced growth lamellae, about 3–4 per mm. Ventral interior with a large, elongate, subtriangular muscle field raised anteriorly, extending almost to mid-valve. Dorsal interior and mantle canals of both valves unknown.

Remarks. Similar small *Eridorthis* are relatively common in the Sandbian deposits of the Chu-Ili and Chingiz-Tarbagatai terranes where they occur in communities characterised by microbial build-ups. They are close to *Eridorthis bestamaki* (Nikitin & Popov *in* Klenina *et al.* 1984), which also has a weakly uniplicate anterior commissure, an almost procline ventral interarea, an anteriorly developed ventral sulcus, and an indistinct dorsal median fold. Their concentric ornament is also very similar. However, the shells from the Berkutsyur Formation differ from *E. bestamaki* in coarser radial ornament, with less than 24 ribs in mature individuals and they are probably not conspecific with it. Specimens of *Eridorthis* sp. (= *Glyptorthis* sp.) from the Anderken Formation of the south Chu-Ili Range (Popov *et al.* 2002) are characterised by a gently sulcate anterior commissure, a dorsal sulcus terminating at the anterior margin, and finer radial ornament with 25–30 costellae, unlike the shells from the Berkutsyur Formation.

Family Plaesiomyidae Schuchert & Cooper, 1931

Genus *Bokotorthis* Popov, Nikitin & Cocks, 2000

Type species (by original designation). – Schizophorella kasachstanica Rukavishnikova, 1956, from the Dulankara Formation (early Katian) of Bokot Well, southern Chu-Ili Range, Kazakhstan.

Bokotorthis kasachstanica (Rukavishnikova, 1956)

Plate 15, figures 1–6, 17

1956   *Schizophorella kasachstanica* sp. nov. Rukavishnikova, p. 118, pl. 1, figs 3–4.

2000   *Bokotorthis kasachstanica* (Rukavishnikova) Popov, Nikitin & Cocks, p. 848, pl. 2, figs 11–18.

Holotype. – IGNA 1369/3, complete shell; Dulankara Formation, Otar Beds (early Katian), from about 3 km south-east of the Bokot Well, southern Chu-Ili Range.

Material. – Kopkurgan Formation, Locality 735: exfoliated ventral valve BC 63818a, ventral internal mould, BC 63846a (Pl. 15, fig. 1); dorsal external moulds, BC 63830a, BC 63847a (Pl. 15, fig. 4), BC 63876a (Pl. 15, fig. 3); BC 63885d; dorsal internal mould, BC 63828 (Pl. 15, fig. 2); dorsal internal moulds, BC 63877 (Pl. 15, fig 5), BC 63882b. Locality 1501: one ventral external mould, NMW 98.28G.2064 (Pl. 15, fig. 17).

Remarks. – This species was originally assigned to *Schizophorella* Reed, 1917, which is a plectorthid, but subsequently designated as the type species of the dinorthid genus *Bokotorthis* after the revision by Popov *et al.* (2002). The specimens from the Kopkurgan Formation show no difference from the topotypes in shell size, radial ornament, characteristic ventral muscle field, and cardinalia, as well as in their well-defined ventral sulcus and a dorsal median fold, and their specific attribution is in no doubt despite their slightly different ages.

Some specimens of *Bokotorthis kasachstanica* (e.g. Pl. 15, fig. 4) shows signs of exopunctae, including rows of aditicules on the rib crescents and occasional epipunctae on the rib flanks; however, it is impossible to provide more details, since those structures are poorly preserved as casts on external mould on the surface of the siliciclastic rock. Epipunctae were also observed on the shells of *Plaesiomys fidelis* Popov *et al.*, 2000, from the Otar Beds (lower Katian) of the Dulankara Mountains in South Kazakhstan. The latter often occur in the rock as the shells have well preserved fibrous structure of the secondary layer. According to Jin *et al.* (2007), Jin & Zhan (2008), and Sproat & Jin (2013), aditicules and epipunctae are setigerous perforations, which was confirmed by the presence of phosphatised setae also with them on the shells of *Plaesiomys* from the Ellis Bay Formation at Anticosti Island. Data from Kazakhstan appears to confirm that such setigerous perforations are present within the whole family.

Superfamily Plectorthoidea Schuchert *in* Schuchert & LeVene, 1929
Family Plectorthidae Schuchert *in* Schuchert & LeVene, 1929

Genus *Altynorthis* n. gen.

Derivation of name. – After altyn, Kazakh for 'golden', due to the location of the type locality near the Akbakai gold mine of one of the species.

Type species. – *Altynorthis vinogradovae* n. sp. from the Berkutsyur Formation (early Sandbian), West Balkhash, southern Central Kazakhstan.

Diagnosis. – Shell subequally biconvex to slightly ventribiconvex, subquadrate to subpentagonal, with hinge line equal or slightly shorter than maximum shell width. Anterior commissure rectimarginate. Ornament of coarsely costate to ramicostellate fine filae. Ventral valve with erect posteriorly pointed beak and high, almost planar interarea. Dorsal valve with variably developed weak median sulcus disappearing towards the mid-length. Ventral interior with suboval to triangular muscle field extended slightly beyond the delthyrial cavity floor. Ventral adductor scar raised anteriorly equal or only slightly narrower and longer than diductor scars. Ventral mantle canals strongly impressed, saccate with *vascula media* curved inwards in front of adductor scars, then divergent distally. Pedicle callist absent. Dorsal interior with a simple blade-like cardinal process on a high notothyrial platform, blade-like brachiophores divergent anteriorly with supports convergent towards the valve floor, and apocopate mantle canals. Fulcral plates well developed. Dorsal anterior adductor scars slightly smaller than the posterior ones.

Species assigned. – In addition to the type species, the genus includes *Altynorthis betpakdalensis* n. sp. from the Baigara Formation (late Darriwilian – early Sandbian), southern Betpak-Dala, Kazakhstan; *Hesperorthis tabylgatensis* Misius, 1986, from the Tabylgaty Formation (Sandbian, *Nemagraptus gracilis* Biozone) of the Moldo-To Range in Kyrgyzstan; *Plectorthis akzharica* Nikitin and Popov 1983, from the Isobilnaya Formation (late Darriwilian) of the Selety River Basin, Kazakhstan; and *Plectorthis numerosa* Nikitin and Popov 1983, from the Kupriyanovka

Formation (early Sandbian) of the Ishim River basin, Kazakhstan.

Species questionably assigned. – *Hesperorthis kara-adirensis* Klenina *in* Klenina *et al.*, 1984, from the Abai Formation (early Sandbian) in the Chingiz Range, Kazakhstan, since there is no adequate data on the ventral valve interior and the mantle canal systems of both valves in the original publication.

Remarks. – There are a number of coarsely ribbed plectorthid species which were previously provisionally assigned to *Plectorthis* due to inadequate knowledge of the type species, *Orthis plicatella* Hall, 1847. Thanks to Sproat & Jin (2016), who completed an extensive revision of that species, including the types originally described and illustrated by Hall (1847), as well as providing a revised generic diagnosis, the taxonomical affiliation of the Kazakh plectorthids can now be resolved. There is a distinct group of species including *Plectorthis akzharica* Nikitin & Popov, 1983, *Plectorthis numerosa* Nikitin & Popov, 1983, and probably closely allied species described as *Hesperorthis karaadirensis* Klenina *in* Klenina *et al.*, 1984 and *Hesperorthis tabylgatensis* Misius, 1986. As noted by Sproat & Jin (2016), those taxa cannot be assigned to *Plectorthis* because of differences in the ventral muscle field morphology. Indeed, while *Orthis plicatella* is characterised by the cordate ventral muscle field typical of most plectorthids (Williams & Harper, 2000), the Kazakh species have a subtriangular or suboval ventral muscle field with a relatively wide and anteriorly raised adductor muscle scar. This justifies the creation of a new genus for those species, including the new species described here, *Altynorthis betpakdalensis* and *A. vinogradovae*. In addition to those features in the ventral muscle field, *Altynorthis* differs from *Plectorthis* in the absence of aditicules, and *Altynorthis* has strongly-impressed mantle canals with distinctive, strongly impressed ventral *vascula media* curved inwards in front of ventral adductor muscle scars which are proximally and divergent distally, as well as apocopate dorsal mantle canals. In most plectorthid genera, the mantle canals are weakly impressed and therefore poorly known. However, in its dorsal cardinalia with a blade-like cardinal process on the high notothyrial platform and high, blade-like brachiophores with convergent supports down to the notothyrial platform, as well as dorsal adductor muscle field with larger posterior scars, *Altynorthis* exhibits typical plectorthid features. The only comparable plectorthid genus is *Desmorthis* Ulrich & Cooper, 1936, but *Altynorthis* can be easily distinguished from *Desmorthis* in having costate radial ornament with considerably coarser ribs, a suboval to subtriangular muscle field with a broader, anteriorly raised ventral adductor scar, and in the absence of an apical plate on the delthyrium.

Altynorthis betpakdalensis n. sp.

Plate 17, figures 13–19; Plate 18, figures 1–8

Holotype. – NMW 98.28G.1469 (Pl. 18, figs 1–3; L, 14.6, W, 14.6, T, 4.8), articulated shell, from Locality 1021 in the Baigara Formation (late Darriwilian to early Sandbian), area *c.* 6 km SW of Baigara Mountain, south Betpak-Dala.

Material. – Baigara Formation: Locality 1020a: two articulated shells, NMW 98.28G.1470 (Pl. 18, figs 5–7), 1724. Locality 1021: 60 articulated shells and one dorsal valve, including paratypes NMW 98.28G.1466 (Pl. 17, fig. 17), 1467 (Pl. 17, fig. 18), 1468 (Pl. 17, figs 13–15), 1469. 1481–1718; Locality 1022: 239 articulated shells, including NMW 98.28G.1472 (Pl. 18, fig. 8), 1473–1480, 1719–1822, 1848–1853. Locality 1023: 153 articulated shells, including NMW 98.28G.1464 (Pl. 17, figs 16, 19), 1465 (Pl. 17, figs 16, 19); dorsal valve NMW 98.28G.1471 (Pl. 18, fig. 4). Locality 1025, 6 articulated shells. Average sizes for 235 articulated shells from Locality1022. Length: X, 13.7; S, 2.87; min, 6.2; max, 21.8. Width: X, 15.3; S, 3.06; min, 7.3; max, 24.8. Thickness: X, 5.25; S, 1.10; min, 2.6; max, 11.4.

Diagnosis. – Slightly ventribiconvex, subquadrate shells, cardinal extremities from obtuse to slightly acute; ventral valve sagittal profile with maximum height anterior to the umbo; radial ornament costate with 20-22 ribs; dorsal sulcus usually absent.

Description. – Shell ventribiconvex, subquadrate on average 92.4% (S, 26.2; OR, 45.1–236.4%; N, 235) as long as wide with maximum width at or slightly anterior to the hinge line and 39.9% (S, 11.3; OR, 19.1–101.6%; N, 235) as thick as long. Cardinal extremities from slightly obtuse to slightly acute, usually almost at right angles. Lateral margins from gently convex to almost straight; anterior commissure rectimarginate, broadly rounded. Ventral sagittal profile moderately convex with maximum height varying between the umbo and the posterior one-third valve length. Ventral beak pointed, posteriorly directed. Ventral interarea high, almost planar to very gently curved, apsacline, with open, broad,

Fig. 27. **A–L.** *Eoanastrophia kurdaica* Sapelnikov & Rukavishnikova, 1975; Berkutsyur Formation (early Sandbian); **A–D,** Locality 8133, BC 65404, ventral, dorsal, lateral and posterior views of conjoined valves; **E–H,** Locality 8124, BC 65403, dorsal, lateral and anterior views of conjoined valves; **I–L,** Locality 8124, BC 65401, posterior, ventral, dorsal and lateral views of conjoined valves. **M–O.** *Paraoligorhyncha?* sp.; Berkutsyur Formation (early Sandbian), Locality 8121; NMW 98.28G.1996, posterior, dorsal and ventral views of conjoined valves. Scale bars 1 mm. **P–Q.** *Lydirhyncha tarimensis* (Sproat & Zhan, 2018); Kopkurgan Formation (Katian), Locality 735; **P,** BC 63872, dorsal internal mould; **Q,** BC 63881, dorsal exterior. Scale bars 2 mm.

triangular delthyrium. Dorsal valve gently convex with maximum height at the umbo. Dorsal interarea low, planar, anacline. Radial ornament coarsely costate with 18–24 rounded ribs (numbering 16–27 in 1, 4, 18, 30, 53, 55, 35, 30, 16, 4, 5, 1 specimens); shells with 20–22 ribs comprising more than 64% of individuals. A few growth lamellae may be present in the anterior half of the shell. Ventral valve with teeth

supported by thin divergent dental plates. Ventral muscle field subtriangular, slightly extending anteriorly beyond the delthyrial cavity floor for about two-fifths valve length. Dorsal adductor scar narrow, subtriangular, raised anteriorly, slightly longer and narrower than gently impressed diductor scars. No distinct pedicle callist. Ventral mantle canals saccate with broad and short *vascula media* slightly curved inward proximally and convergent distally. Dorsal interior with a simple, ridge-like cardinal process on a high notothyrial platform. Brachiophores blade-like, divergent, with supporting plates convergent dorsally towards the notothyrial platform. Adductor muscle scars gently impressed with a slightly larger posterior pair, bisected by a broad median ridge. Dorsal mantle canals apocopate.

Remarks. – While almost all the shells are costate, about 1% of them show bifurcations within one or two ribs in the mid part of the shell. A weakly-defined dorsal sulcus originating at the umbo and fading anteriorly is present in fewer than 5% of individuals. A cluster of two articulated shells (Pl. 17, figs 16, 19) closely attached to each other and probably in life position suggests that *Altynorthis betpakdalensis* had a very short pedicle; perhaps a pad with short rootlets emerging from the delthyrial opening, which explains the absence of a distinct pedicle callist in the umbonal part of the delthyrial cavity floor. The new species has superficial similarity to *Altynorthis numerosa* (Nikitin & Popov, 1983), from the Kupriyanovka Formation (early Sandbian) of the Ishim Region in northern Kazakhstan (the Kalmykkol-Kokchetav Terrane) in having a slightly more transverse, subrectangular shell, often with rectangular and acute cardinal extremities, an uneven sagittal profile of both valves with maximum height near the umbonal area, high and almost planar ventral interarea, strongly pointed, posteriorly erect (not curved) ventral beak, radial ornament with a few rib bifurcations occurring in about 1% of individuals, and strongly impressed, raised anteriorly ventral adductor scars. *Altynorthis akzharica* (Nikitin & Popov, 1983) from the Izobilnaya Formation (late Darriwilian) of the Selety Terrane is comparable with *Altynorthis betpakdalensis* in its uneven sagittal profile of both valves, posteriorly erect ventral beak and high, almost planar ventral interarea, but the latter species can be readily distinguished from the former in its coarser radial ornament with fewer ribs (18–24 against 26–32), and only sporadic development of a very weakly defined dorsal sulcus. Also the ventral *vascula media* in *Altynorthis akzharica* are almost straight and anteriorly divergent through their entire length, while in *A. betpakdalensis* they are distinctly curved inwards (Pl. 17, fig. 18). *Altynorthis betpakdalensis* differs from the type species *A. vinogradovae* mainly in having costate, not costellate radial ornament and a slightly larger ventral muscle field, while other features of the external and internal shell morphology are very similar.

Altynorthis tabylgatensis (Misius, 1986)

Plate 18, figures 9–14; Figure 24L

1986 *Hesperorthis tabylgatensis* sp. nov. Misius, p. 104, pl. 1, figs 1–26, pl. 2, figs 1–4.

Holotype. – Geological Institute, Bishkek, Kyrgyzstan 14/5, articulated shell, from the Tabylgaty Formation (Sandbian, *Nemagraptus gracilis* Biozone), Locality 317–318, left-hand side of Tabylgaty River, Moldo-Too Range, north Kyrgyzstan (North Tien-Shan Terrane).

Material. – Baigara Formation: Locality 1026: three ventral and six dorsal valves preserved as internal and external moulds, including BC 60833 (Pl. 18, figs 9, 10) and ventral internal mould NMW 98.28G.1973 (Fig. 24L). Locality 1026b: seven articulated shells and one dorsal valve including NMW 98.28G.1461, 1973–1979. Locality 1028: 11 ventral and nine dorsal valves preserved as internal and external moulds, including, NMW 98.28G.1238; dorsal internal mould, NMW 98.28G.1238 (Pl. 18, fig. 11). Berkutsyur Formation, Locality 8234: two ventral internal moulds, BC 63866 (Pl. 18, figs 12, 13), BC 62349b. Takursu Formation, Locality 166, *c.* 50 disarticulated valves and shell fragments, including NMW 98.28G.2060 (Pl. 18, fig. 14).

Remarks. – *Altynorthis tabylgatensis* is characterised by a slightly ventribiconvex shell, coarsely costate radial ornament with 16–17 ribs, ventral sagittal profile with maximum height in the umbonal area, high, apsacline ventral interarea, a subtriangular ventral muscle field, dorsal cardinalia with a blade-like cardinal process on high notothyrial platform, blade-like brachiophores with support convergent towards notothyrial platform, and saccate ventral and apocopate dorsal mantle canals. All these features are characteristic of *Altynorthis* and also for the specimens from the Baigara Formation collected in the area of Karatal River in south Betpak Dala: the only major difference of the latter is in their wide range of ribs, which are up to 19 in some specimens. Although the presence of fulcral plates was not reported in the Kyrgyz specimens, and while details of mantle canal system were not seen on the specimens from Karatal, there is nevertheless enough evidence to consider them as conspecific. *Altynorthis tabylgatensis* differs from *A. betpakdalensis* in its coarser radial ornament with 16–19 ribs, which is mainly below the limits observed on the shells of the latter taxon, and in the presence of a shallow yet well-defined dorsal sulcus: also individuals with a hinge line shorter than maximum shell width are the most common.

Altynorthis vinogradovae n. sp.

Plate 19, figures 1–12; Figure 25A

Derivation of name. – After the late Elena V. Vinogradova in appreciation of her studies on the Early Palaeozoic geology of the West Balkhash region.

Holotype. – BC 62356 (Pl. 19, figs 5, 6), dorsal internal mould, Locality 814, Berkutsyur Formation (early Sandbian), area 4 km south-west of Lake Alakol, West Balkhash.

Paratypes. – Berkutsyur Formation: Locality 812: one articulated shell two ventral and one dorsal valves, including BC 60802-03. Locality 813: three articulated shells, eight ventral and 17 dorsal valves, including BC 60794-99, BC 60800 (Lv, 19.4; Ml, 8.4; Mw, 7.6), BC 60826 (Lv, 20.9; Ml, 5.4; Mw, 5.9; Pl. 19, fig. 4), BC 60827 (Lv, 14.6; W, 21.5; Fig. 25A), BC 60828 (Pl. 19, fig. 7); Ld, 16.2; W, 18.4), BC 62345 (Pl. 19, figs 2, 3), BC 62346 (Pl. 19, fig. 11), BC 62347 (Pl. 19, fig. 8), BC 62348 (Pl. 19, fig. 12). Locality 814: four articulated shells, seven ventral and four dorsal valves, including BC 60801 (Pl. 19, fig. 1), BC 60805-8, BC 60832, BC 62357 (Pl. 19, fig. 10), BC 62358a, BC 62359, NMW 98.28G.1971, 1972 (Pl. 19, fig. 9). Locality 815: one dorsal valve, BC 60804. Locality 816: one articulated shell. Locality 8120: one dorsal valve, BC 60831. Locality 8124: one ventral and 3 dorsal valves). Locality 8125; two ventral internal moulds, BC 60829a, BC 60830, and one dorsal internal mould, BC 60829b. Locality 8233: conjoined valves, BC 60814 (Lv, 21.8; Ld, 21.6; W, 26.7; T, 8.8).

Diagnosis. – Shell slightly ventribiconvex, subpentagonal, cardinal extremities slightly obtuse; ventral valve sagittal profile with maximum height in front of the umbo; radial ornament ramicostellate.

Description. – Shell slightly ventribiconvex, subpentagonal outline, about three-quarters as long as wide with maximum width slightly anterior to the hinge line and about 20% as thick as long. Cardinal extremities slightly obtuse; anterior commissure

rectimarginate. Radial ornament ramicostellate with 18-20 ribs at the umbo, up to 50 rounded ribs along the anterior and lateral shell margins, and 4–5 ribs per 5 mm along the anterior margin of mature individuals. Concentric ornament of faint, ridge-like, densely crowded filae about 4 per mm. Ventral valve sagittal profile moderately convex with maximum height in front of the umbonal area and very gently convex anterior slope. Ventral beak pointed, posteriorly erect. Ventral interarea high, apsacline, almost planar, with open narrow triangular delthyrium. Dorsal valve moderately convex with maximum height slightly posterior to mid-length. Dorsal interarea low planar, anacline with an open notothyrium. Ventral interior with teeth supported by short dental plates, thickened at the base, and slightly elongate, triangular muscle field about one-third valve length. Ventral adductor scar gently impressed narrow, triangular, separated laterally by a pair of faint myophragms from slightly wider diductor scars of equal length. Ventral mantle canals saccate with strongly impressed *vascula media* gently curved inward in front of the adductor scars, then divergent distally (Fig. 25A). Dorsal interior with a simple blade-like cardinal process bisecting a high, subtriangular, slightly inclined posteriorly notothyrial platform and blade-like, anteriorly divergent, brachiophores with supports convergent downwards towards the notothyrial platform (Pl. 20, figs 5, 11). Adductor scars gently impressed with a slightly larger posterior pair bisected by a broad, low median ridge. Dorsal mantle canals apocopate.

Remarks. – *Altynorthis vinogradovae* is the only known species within the genus with ramicostellate radial ornament. While a few occasional rib bifurcations occur in other species, they all have costate radial ornament.

Genus *Lictorthis* n. gen.

Derivation of the name. – Latin *lictus*, abandoned and *Orthis*, much quoted brachiopod generic name.

Type and only species. – *Plectorthis licta* Popov & Cocks, 2006, from the Degeres Member of the Dulankara Formation (mid Katian), southern Chu-Ili Range, south Kazakhstan.

Diagnosis. – Shell subequally biconvex subrectangular, with a hinge line slightly shorter than maximum width at mid length and obtuse cardinal extremities;

anterior commissure rectimarginate. Ventral valve gently convex, posteriorly subcarinate, with a low, gently curved, apsacline interarea and open delthyrium. Dorsal valve with planar orthocline interarea and shallow sulcus originating at the umbo and gradually fading forwards. Ornament multicostellate with fine filae. Ventral interior with triangular muscle field. Adductor scar strongly raised anteriorly about equal length and width with the diductor scars. Dorsal mantle canals saccate with thin, widely divergent vascula *media*. Dorsal interior with a notothyrial platform bisected by the blade-like cardinal process bearing crenulated myophore. Brachiophores high, blade-like with supporting plates converging down onto the notothyrial platform. Deep sockets confined anterolaterally by well developed fulcral plates. Adductor scars small, strongly impressed with a larger posterior pair. Mantle canals lemniscate.

Remarks. – As well as *Lictorthis*, there are two endemic Ordovician plectorthide genera *Altynorthis* and *Weberorthis* in the Kazakh terranes, all of which have conservative plectorthid morphology of their cardinalia, with the only differences in the presence or absence of crenulation of the myophore and the lack of a pedicle callist. However, they are unlike peri-Iapetus and Laurentian plectorthid genera with a cordate ventral muscle field which all show a less derived, triangular or suboval muscle field with a prominent adductor scar raised anteriorly and of about equal length with the diductor scars (Williams & Harper 2000). They also lack aditicules, unlike *Plectorthis*. This suggests that the Kazakh plectorthids may represent an early offshoot of the family which separated sometime in the Darriwilian and then evolved in relative geographical isolation. In comparison with the two other Kazakh genera of the family, *Lictorthis* can be distinguished from *Altynorthis* in having a larger, gently biconvex shell with multicostellate radial ornament, thin widely divergent ventral *vascula media*, lemniscate dorsal mantle canals and a cardinal process with crenulated myophore. It also has coarser radial ornament, a rectimarginate anterior commissure and lacks a prominent dorsal median fold and ventral sulcus (unlike *Weberorthis*).

Lictorthis cf. *L. licta* (Popov & Cocks, 2006)

Plate 15, figures 9, 10

cf. 2006 *Plectorthis licta* sp. nov. Popov & Cocks, p. 276, pl. 6, figs 10, 13-17, text-fig. 6F.

Remarks. – This taxon is represented by a single ventral internal mould, BC 63874b, from Locality 735 in the Kopkorgan Formation, which exhibits features characteristic of *Plectorthis licta* in the ventral interarea, muscle field and mantle canals, and there is no other Late Ordovician orthoid taxon with which it can be compared. However, the species assignment remains provisional because of the sparse material.

Family Giraldiellidae Williams & Harper, 2000

Genus *Phaceloorthis* Percival, 1991

Type species (by original designation). – *Phaceloorthis decoris* Percival, 1991, from the Qondong Limestone (Katian), New South Wales, Australia.

Phaceloorthis? corrugata n. sp.

Figure 26A–K

Derivation of name. – After the characteristic corrugate concentric ornament in the anterior part of the shell.

Holotype. – BC 63876b (Figs 26A, B), dorsal internal mould, from Locality 735, Kopkorgan Formation (middle Katian), area 10 km south-west of Lake Alakol, West Balkhash.

Topotypes. – Locality 735: six ventral and three dorsal valves preserved as internal and external moulds, including BC 63875 (Fig. 26J), BC 63820c (Figs 26 G–I), BC 63820d (Fig. 26F), BC 63821a (Figs 26C, D), BC 63838c, BC 63839b (Fig. 26K), BC 63883 (Fig. 26E).

Diagnosis. – Shell subequally biconvex with rectimarginate anterior commissure, dorsal sulcus only in the umbonal area, radial ornament coarsely fascicostellate with 10–14 primary ribs at the umbo; ventral muscle field cordate with narrow adductor scars shorter than diductor scars.

Description. – Shell subequally biconvex, transverse, subrectangular outline, about three-quarters as long as wide, with a maximum width at about mid-length. Cardinal extremities obtuse. Anterior commissure rectimarginate. Ventral valve sagittal profile moderately convex with maximum height between the umbo and mid-length. Ventral muscle field about two-fifths valve length. Ventral interarea moderately high, apsacline, gently curved in cross section, with an open triangular delthyrium. Dorsal valve sagittal profile moderately convex with maximum height at one-third valve length. Dorsal interarea low, planar, strongly anacline with an open notothyrium. A very shallow sulcus present in the umbonal area, but fading completely posteriorly to mid-valve. Radial ornament fascicostellate with 10–14 primary ribs at the umbo and up to 60 ribs along the anterior and lateral margins of large individuals and with up to 7 ribs per 5 mm along the posterior margin. Concentric ornament with regularly spaced ridge-like fila about 4–5 per mm in the mid part of the shell, becoming more pronounced and corrugate closer to the anterior margin, where their density decreases to 2–3 per mm (Figs 26G, K), where numerous growth lamellae also appear. Ventral interior with teeth supported by long, narrowly divergent dental plates extending anteriorly into distinct muscle bounding ridges enclosing the entire ventral muscle field. Ventral muscle field cordate, about two-fifths valve length, extending anteriorly beyond the delthyrial cavity floor. Ventral adductor scar narrow, distinctly shorter, but not enclosed by the longer diductor scars and separated from them laterally by faint myophragms. Pedicle callist not impressed. Ventral mantle canals weakly impressed, a pair of thin, straight anteriorly divergent *vascula media* seen on a single specimen (Fig. 26J). Dorsal interior with a high notothyrial platform bisected by a blade like cardinal process thickened anteriorly and with a crenulated myophore on the posterior face. Brachiophores blade-like, widely divergent, with short bases convergent with the notothyrial platform. Fulcral plates well defined (Figs 26A, B). Ventral adductor muscle field small weakly impressed, bisected by a thick, short median ridge. Individual muscle scars indiscernible. Mantle canals weakly impressed, probably lemniscate.

Remarks. – There is another species of the genus, *Phaceloorthis recondita* Popov *et al.*, 2000, from the Otar Member of the Dulankara Formation in the southern Chu-Ili Range, Kazakhstan. *Phaceloorthis corrugata* can be distinguished from it, and also from the type species *Phaceloorthis decoris* from Australia, in its much coarser radial ribs (10–14 primary ribs at the umbo), very poorly defined dorsal sulcus completely fading to mid-valve, and a cordate (not

triangular) ventral muscle field with short and narrow adductor scar. The latter feature makes generic assignment of the species somewhat tentative: hence the query.

Genus *Scaphorthis* Cooper, 1956

Type species (by original designation). – *Scaphorthis virginiensis* Cooper, 1956, from the Chatham Hill Formation (Sandbian) of Virginia, USA.

Scaphorthis recurva Nikitina, 1985

Figure 25B–L

1985 *Scaphorthis recurva* sp. nov. Nikitina, p. 23, pl. 1, figs 8–13.

Holotype. – CNIGR 8/12093, Rgaity Formation (late Darriwilian), vicinity of the Talapty winter hut, South Kendyktas Range, North Tien Shan Terrane, Kazakhstan.

Material. – Baigara Formation, Locality 1021: 21 articulated shells, one ventral valve, including NMW 98.28G.1869 (Fig. 25D), 1870 (Figs 25H-J), 1871 (Fig. 25K, L), 1874 (Fig. 25E, F), 1875–77, 2017–2033. Locality 1022: fifteen articulated shells, including NMW 98.28G.1854–1868. Locality 11221: 13 ventral and 15 dorsal valves preserved as internal and external moulds including, NMW 98.28G.2075 (Figs 25B, C), 2065–2074, 2086–2100.

Remarks. – Nikitina (1985) presented a detailed description of *Scaphorthis recurva* and also considered the shells described by Nikitin & Popov *in* Klenina *et al.* (1984) as *Scaphorthis* ex gr. *perplexa* Cooper, 1956, from the lower part of the Bestamak Formation (late Darriwilian – early Sandbian) of the Chingiz Ranges, as conspecific with the newly designated species. However, because of the poor preservation of specimens, as illustrated in the cited publication, their precise specific identification is doubtful and therefore they are not included in the synonymy above. Shells of *Scaphorthis recurva* are common in the lower part of the Baigara Formation, although variably distorted; and direct comparison with the topotypes, also represented in the studied collection (Fig. 25B, C), indicates without doubt that they are conspecific.

Suborder Dalmanellidina Moore, 1952

Superfamily Dalmanelloidea Schuchert, 1913

Family Dalmanellidae Schuchert, 1913

Subfamily Dalmanellinae Schuchert, 1913

Genus *Onniella* Bancroft, 1928

Type species, (by original designation). – *Onniella broeggeri* Bancroft, 1928, from the Acton Scott Formation (Sandbian), Shropshire, England.

Onniella? sp.

Figure 26L–N

Material. – Kopkurgan Formation: Locality 735: ventral external mould, BC 63839c (Fig. 26L); dorsal internal mould, BC 63888c (Fig. 26M, N).

Remarks. – Endopunctate orthides are extremely rare in the Katian of the Chu-Ili Terrane and only an unnamed species of *Epitomyonia* has previously been documented (Popov & Cocks 2006). A dorsal valve interior from the uppermost Kopkurgan Formation shows the characteristic cardinalia with a small, swollen, undifferentiated cardinal process and thin brachiophores with divergent supporting plates. No adductor scars or median ridge are seen. An incomplete ventral exterior provisionally associated with the same unnamed taxon exhibits costellate ornament of subangular ribs, increasing in number by bifurcation, separated by interspaces about the same width. The simple morphology of the cardinalia seems closest to *Onniella*, so the Kazakh shells are provisionally assigned to that genus.

Superfamily Enteletoidea Waagen, 1884

Family Draboviidae Havlíček, 1950

Subfamily Draboviinae Havlíček, 1950

Genus *Pionodema* Foerste, 1912

Type species (by original designation). – *Orthis subaequata* Conrad, 1843, from the Decorah Formation (Katian) of Wisconsin, USA.

Pionodema opima Popov, Cocks & Nikitin, 2002

Plate 16, figures 23–26; Plate 20, figures 3–10

2002 *Pionodema opima* sp. nov. Popov, Cocks & Nikitin, p. 64, pl. 12, figs 13, 14, 16–27.

Holotype. – BC 57545, internal mould of conjoined valves, from the Anderken Formation (Sandbian) of Kujandysai, Chu-Ili Range, southern Kazakhstan.

Material. – Berkutsyar Formation, Locality 817: two articulated shells including one internal mould, BC 62397 (Pl. 20, figs 5–8), BC 62398 (Pl. 20, figs 9, 10). Locality 8235a: one ventral valve, NMW 98.28G.1967 (Pl. 16, figs 25, 26) and two dorsal valves, NMW 98.28G.1968 (Pl. 20, figs 3, 4), 1969 (Pl. 16, fig. 23), 1970 (Pl. 16, fig. 24).

Remarks. – This is the only endopunctate orthide species yet documented from the Sandbian of the Chu-Ili Terrane. The specimens from Locality 817 are large and show an uniplicate anterior commissure, a dorsal median fold and a ventral sulcus characteristic of the species (Pl. 20, figs 5–8) together with well impressed mantle canals in both valves (Pl. 20, figs 9, 10). The specimens from Locality 8235a are juveniles which were probably displaced from their original habitat by a mass flow.

Order Pentamerida Schuchert & Cooper, 1931

Suborder Syntrophiidina Ulrich & Cooper, 1936

Superfamily Camerelloidea Hall & Clarke, 1894

Family Camerellidae Hall & Clarke, 1894

Genus *Ilistrophina* Popov, Cocks & Nikitin, 2002

Type species (by original designation). – *Ilistrophina tesikensis* Popov, Cocks & Nikitin, 2002, from the Anderken Formation (Sandbian) of Tesik River, Chu-Ili Range, southern Kazakhstan.

Ilistrophina tesikensis Popov, Cocks & Nikitin, 2002

Plate 20, figures 12–14, 16

2002 *Ilistrophina tesikensis* sp. nov. Popov, Cocks & Nikitin, p. 69, pl. 12, figs 32–35.

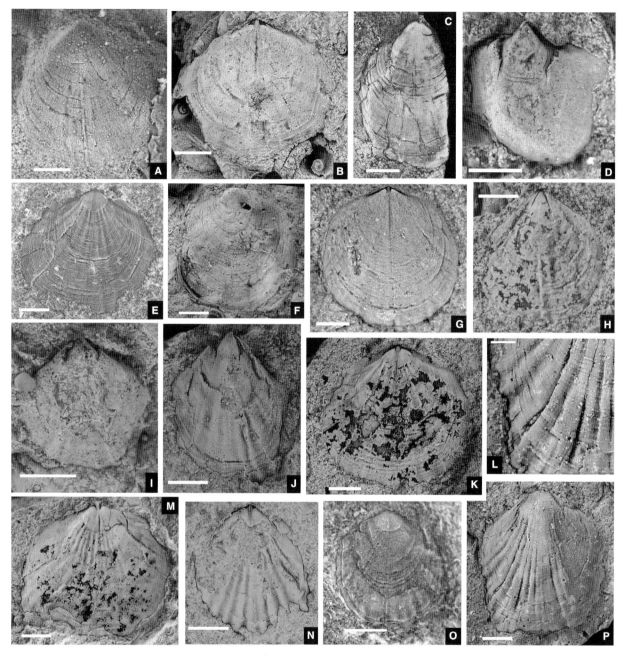

Fig. 28. **A–P.** *Costistriispira proavia* gen. et sp. nov.; Berkutsyur Formation (early Sandbian), Locality 8234; **A,** BC 63869b, latex cast of ventral exterior; **B,** BC 63867a, dorsal internal mould; **C,** BC 63831b, latex cast of ventral exterior; **D,** BC 63817b, ventral internal mould; **E,** BC 63856, latex cast of dorsal exterior; **F,** BC 63852, latex cast of ventral exterior; **G,** BC 63868b, dorsal internal mould; **H,** BC 63817a, ventral internal mould; **I,** BC 63863, ventral internal mould of juvenile shell; **J,** BC 63870a, ventral internal mould; **K,** BC 63864, dorsal internal mould; **L, P,** BC 63869a, detail of radial and concentric ornament, latex cast of ventral exterior; **M,** BC 63892a, holotype, dorsal internal mould; **N,** BC 63870b, dorsal internal mould; **O,** BC 63859a, dorsal exterior; A–K, M–P, scale bars 2 mm; L, scale bar 1 mm.

Holotype. – BC 56823, conjoined valves, from the Anderken Formation (Sandbian) of Tesik River, Chu-Ili Range, southern Kazakhstan.

Material. – Berkutsyur Formation, Locality 8121: dorsal valve BC 63697; ventral valve NMW 98.28G.1995. Locality 8124: articulated shell BC 62423 (Pl. 20, figs 12–14, 16).

Remarks. – No transverse serial sections were made, nevertheless traces of the sessile spondylium posteriorly raised on a low median septum anteriorly, and a dorsal median septum supporting a septalium can be observed on the surface of a slightly exfoliated umbonal part of the shell. Furthermore, the specimens show a uniplicate anterior commissure, a dorsal median fold, and a ventral sulcus well

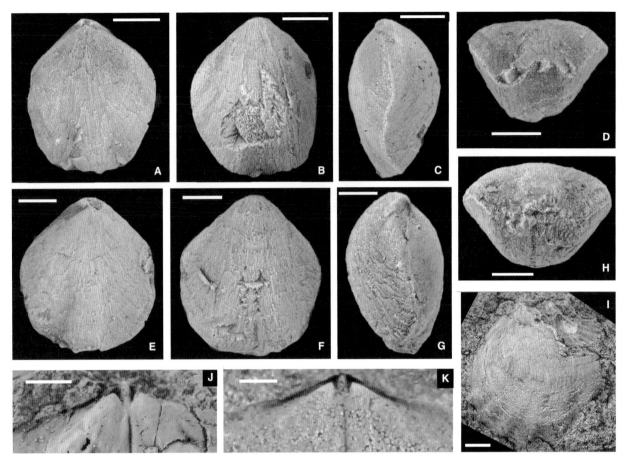

Fig. 29. **A–H.** *Rozmanospira mica* (Nikitin & Popov in Klenina *et al.*, 1984); Berkutsyur Formation (early Sandbian), Locality 8121; **A–D,** NMW 98.28G.1989, dorsal, ventral, side and anterior views of articulated shell; **E–H,** NMW 98.28G.1990, dorsal, ventral, side and anterior views of articulated shell; scale bars 1 mm. **I.** *Parastrophina?* sp., Kopkurgan Formation (late Sandbian), Locality 727; BC 63814, incomplete dorsal valve. Scale bar 2 mm. **J, K.** *Costistriispira proavia* n. gen. et n. sp.; Berkutsyur Formation (early Sandbian), Locality 8234; **J,** BC 63892a, holotype, dorsal internal mould, enlarged cardinalia; **K,** BC 63868b, dorsal internal mould enlarged cardinalia. Scale bars 2 mm.

defined anterior to mid-length. All these features in combination are characteristic of *Ilistrophina tesikensis* and distinguish it from *Liostrophia pravula* Popov *et al.* 2002, which is another smooth camerellid from the Sandbian of the Chu-Ili Terrane.

Genus *Plectocamara* Cooper, 1956

Type species (by original designation). – Plectocamara costata Cooper, 1956, from the Lincolnshire Limestone (Darriwilian), Tennessee, USA.

Plectocamara sp.

Plate 16, figures 17, 22, 27; Figure 23F, G

Material. – Baigara Formation, Locality 765-e: two articulated shells, including NMW 98.28G.1982 (Figs 23F, G). Locality 1021: one articulated shell, including NMW 98.28G.1981 (L, 9.4, W, 8.3, T, 4.9, Sw, 5.9; Pl. 16, figs 17, 22, 27).

Description. – Shell astrophic, subequally biconvex, slightly elongate, subtriangular outline with maximum width at about one-third of anterior margin and about half as thick as long. Anterior commissure strongly uniplicate. Ventral valve lateral profile moderately convex with maximum height at about one-quarter valve length from the umbo. Beak small posteriorly erect. Ventral sulcus originates at the umbo, with steep, flat lateral sides, terminated with the high trapezoidal tongue. Dorsal valve lateral profile moderately convex with maximum height anterior to mid-length. Dorsal median fold high, with flat top and steep lateral sides, originating in the umbonal area some distance from the umbo. Radial ornament of strong angular ribs with three ribs in the ventral sulcus, four ribs on the dorsal median fold and 6–8 on flanks. A single rib bifurcates in the sulcus and dorsal median fold anterior to mid-length. Ventral interior with a spondylium simplex supported by a thin median septum. Dorsal interior with discrete, subparallel inner hinge plates attached to the valve floor. No alate plates.

Remarks. – In their radial ornament the Kazakh shells are most similar to *Plectocamara aseptata* Cooper, 1956, from the Sandbian of eastern USA, but differ in their considerably larger shell size, the presence of a secondary rib inclined in the mid part of the ventral sulcus and dorsal median fold, and a broad sulcus which is well defined in the umbonal area and anteriorly. These shells probably represent a new species, but it cannot yet be named due to the poor preservation and strong distortion of the shells.

Family Parastrophinidae Schuchert, 1929

Genus *Eoanastrophia* Nikiforova & Sapelnikov, 1973

[= *Kokomerena* Misius, 1986, p. 182)]

Type species (by original designation). – *Eoanastrophia antiquata* Nikiforova & Sapelnikov, 1973, from the Archalyk Beds (upper Katian) of the Zeravshan Range, Uzbekistan.

Remarks. – The diagnosis of *Eoanastrophia* given by Sapelnikov & Rukavishnikova (1975) states that the inner hinge plates in *Eoanastrophia* are separate and do not merge with the valve floor to form a septalium. Also that the ventral valve interior is characterised by a spondylium duplex which is sessile posteriorly and supported by a median septum anteriorly which can be longer or shorter than the spondylium. In both attributes, the diagnosis of the genus given by Carlson (*in* Kaesler 2002, p. 955) is erroneous. *Kokomerena* Misius, 1986 is indistinguishable from *Eoanastrophia*, as discussed below, and should be considered a junior synonym of it.

Eoanastrophia kurdaica Sapelnikov & Rukavishnikova, 1975

Figure 27A–L

1975 *Eoanastrophia kurdaica* sp. nov. Sapelnikov & Rukavishnikova, p. 34, pl. 19, figs 10–12, text-fig. 6.

1986 *Kokomerena prima* Misius p. 182, pl. 21, figs 10–22.

Holotype. – IGNA 148/1861, a pair of conjoined valves from the Keskentas Formation (Sandbian) of Kendyktas Range, North Tien Shan, Kazakhstan.

Material. – Berkutsyur Formation, Locality 8120/4b: one articulated shell. Locality 8121: one articulated shell NMW 98.28G.1948. Locality 8124: four articulated shells, including BC 65401 (Fig. 27I–L), BC 65402, BC 65403 (Fig. 27E–H). Locality 8233: seven articulated shells and one ventral and one dorsal valve, including, BC 65404 (Fig. 27A–D), BC 65405-11, NMW 98.28G.1946–47.

Remarks. – *Eoanastrophia kurdaica* was previously known only from the Sandbian of the North Tien Shan Microcontinent, so this is the first report of its occurrence in Chu-Ili. The taxon was described by Sapelnikov & Rukavishnikova, 1975, who also gave detailed sketches of the interiors of both valves. Previously published half-tone illustrations of the taxon are poor and thus new photographs of the specimens from Chu-Ili are provided here. The type and only species of *Kokomerena*, which is *K. prima* Misius, 1986, is a junior synonym of *Eoanastrophia kurdaica* Sapelnikov & Rukavishnikova, 1975. Both are characterised by slightly dorsibiconvex shell with strongly uniplicate anterior commissure, ventral sulcus and dorsal median fold originating at the posterior part of the shell and coarse radial ornament of subangular ribs with 3–4 ribs in the sulcus, 4–5 ribs on the median fold and 6–8 on the flanks of the shell. The transverse serial sections in Misius (1986) are almost identical with those in Sapelnikov & Rukavishnikova (1975, p. 6), and both show a deep, bell-shaped spondylium duplex sessile posteriorly which is supported by the median septum anteriorly, and thin, dorsally convergent inner hinge plates which are not united on the shell floor. The only difference is that the faint alate plates shown by Sapelnikov & Rukavishnikova (1975) were not spotted by Misius (1986).

Genus *Parastrophina* Nikiforova & Sapelnikov, 1973

Type species (by original designation). – *Atrypa hemiplicata* Hall, 1847, from the Martinsburg Formation (Darriwilian), Virginia, USA.

Parastrophina? sp.

Figure 29I

Remarks. – A single incomplete dorsal valve, BC 63814 (Fig. 29I) from Locality 727 in the Berkutsyur Formation, shows paucicostate radial ornament of four broad, rounded ribs originating in the anterior part of the shell and a distorted median fold originated near mid-length. No ventral valves or dorsal valve interior are known. The specimen recalls *Parastrophina iliana* Popov et al., 2002, which is relatively common in carbonates of the Anderken Formation (Sandbian) associated with bacterial build-ups in the southern Chu-Ili Range. Another species of *Parastrophina* which occurs in the Sandbian of Chu-Ili is *P. plena* Sapelnikov & Rukavishnikova, 1975, but that completely lacks ribs on its flanks. Whilst generic assignment of the specimen is probable, there is insufficient material to erect a new species.

Order Rhynchonellida Kuhn, 1949
Superfamily Ancistrorhynchoidea Cooper, 1956
Family Ancistrorhynchidae Cooper, 1956

Genus *Ancistrorhyncha* Ulrich & Cooper, 1936

Type species (by original designation). – *Ancistrorhyncha costata* Ulrich & Cooper, 1936, from the Bromide Formation (Sandbian) of Oklahoma, USA.

Ancistrorhyncha modesta Popov *in* Nikiforova & Popov, 1981

Plate 20, figures 11, 15, 17–19

1981 *Ancistrorhyncha modesta* Popov *in* Nikiforova & Popov, p. 61, pl. 7, figs 6–9, pl. 8, figs 2, 3.

Holotype. – CNIGR1/11774, dorsal valve from Locality 1020, Baigara Formation (late Darriwilian to early Sandbian), area about 6 km SW of Baigara Mountain, south Betpak-Dala, South Kazakhstan.

Material. – Baigara Formation, Locality 1020, one articulated shell, 43 ventral and 43 dorsal valves, including CNIGR 1-7/11774. Berkutsyur Formation, Locality N-12: seven ventral and nine dorsal valves, including BC 60755-63, NMW 98.28G. 2007 (Pl. 20, fig. 17), 2008 (Pl. 20, fig. 18), 2009 (Pl. 20, figs 11, 15), 2016 (Pl. 20, fig. 19). Total one articulated shell, 50 ventral and 52 dorsal valves.

Remarks. – *Ancistrorhyncha modesta* is probably the earliest rhynchonellide species in the Kazakh terranes, and was the core of the nearshore low diversity rhynchonellide-dominated community described by Nikiforova & Popov (1981), so there is no need to repeat the details here. However, in addition to the type locality in the area south-west of Baigara Mountain in the lowermost part of the Baigara Formation, the species is also found in the lower part of the Berkutsyur Formation in West Balkhash and some specimens are illustrated here.

Genus *Lydirhyncha* n. gen.

1956 *Rhynchotrema* (*pars*) Rukavishnikova, [*R. otarica* only], p. 156.

1964 *Rhynchotrema* (*pars*) Wang *in* Wang & Jin, [*R. zhejiangensis* only], p. 46.

1978 *Altaethyrella* Severgina, p. 38.

1983 *Rhynchotrema* (*pars*) Wang *in* Wang & Jin, [*R. zhejiangensis* only], p. 46.

1998 *Altaethyrella* Zhan & Li, [*R. zhejiangensis* only], p. 435,

1998 *Altaethyrella* Zhan & Cocks, p. 54.

2018 *Rhynchotrema* (*pars*) Sproat & Zhan, [*R. tarimensis* only], p. 46.

Derivation of name. – After the late Lydia M. Paletz in appreciation of her valuable studies in the Ordovician and Silurian geology of the West Balkhash Region.

Type species. – *Rhynchotrema zhejiangensis* Wang *in* Wang & Jin, 1964, from the upper Katian of South China.

Diagnosis. – Shell dorsibiconvex, slightly transverse to almost equidimensional, with a well-developed dorsal median fold and ventral sulcus originating in the umbonal area. Radial ornament of strong, simple, subangular ribs originating at the umbo. Ventral interior with vestigial dental plates, a narrow, strongly elongate ventral muscle field with an adductor scar completely separating strongly impressed diductor muscle scars. Dorsal interior with short, disjunced hinge plates and a faint, septiform cardinal process. No cruralium supported by the dorsal median septum.

Species assigned. – In addition to the type species, *Lydirhyncha* includes *Rhynchotrema gushanensis* Liang *in* Liu *et al.*, 1983, a junior synonym of *Lydirhyncha zhejiangensis* according to Zhan & Cocks (1998), from the Xiazhen Formation and upper Changwu formations (late Katian) of South China, *Rhynchotrema tarimensis* Sproat & Zhan, 2018, from the Hadabulaktag Formation (Katian) of the Kuruktag region at Xinjiang, north-west China, and *Rhynchotrema yaoxianensis*, Fu, 1982, from the Jinhe Formation (middle to late Katian) of North China.

Species questionably assigned. – *Latirhyncha inflata* Xu, 1982, from the Shiyanhe Formation (Katian) of North China; *Drepanorhyncha pentagonia* Fu, 1982, and *Drepanorhycha triplicata* Fu, 1982, both from the Jinhe Formation (middle to late Katian) of North China.

Discussion. – The taxonomy of Kazakh Late Ordovician rhynchonellides is a matter of much confusion (Savage 2006, p. 2724; Sproat & Zhan 2018). *Rhynchotrema otarica*, as originally described by Rukavishnikova (1956), is probably a heterogenous taxon. Nikiforova & Popov (1981), during their revision of the Upper Ordovician rhynchonellides, discovered that the rhynchonellide shells from almost all the localities (including Dulankara and Zhartas) listed by Rukavishnikova (1956) lack a cruralium and median septum and therefore concluded that the species should be transferred to the family Ancistrorhynchidae, and thus erected the genus *Otarorhyncha* to accommodate it. However, in transverse sections of the shell from the type locality for *Rhynchotrema otarica* (Rukavishnikova 1956, text-fig. 5), the area near the Bokot well, a cruralium supported by a low median septum is clearly visible, as pointed out by Savage (2002), who reassigned *Otarorhyncha* to the family Rhynchotrematidae. However, in external shell morphology those shells are indistinguishable from other Dulankara rhynchonellides. Apart from *Rostricellula sarysuica*

Nikitin *et al.*, 1996, there are no Katian rhynchonellides with a cruralium and dorsal median ridge from South Kazakhstan (Popov *et al.* 2000; Popov & Cocks 2006), including the new locality from the Kopkurgan Formation of West Balkhash. However, the presence or absence of a cruralium in the rhynchonellide shells from the Bokot well cannot be verified from new topotype material since no precise location was given for that well in the original description by Rukavishnikova (1956) of '*Rhynchotrema*' *otarica* and the locality is absent on all the available topographical maps of the south Chu-Ili Range. The only solution to resolve this controversy, is to restrict the binomen '*Rhynchotrema*' *otarica* only to the shells collected from the Bokot well and illustrated by Rukavishnikova (1956, pl. 5, figs 7, 8, text-fig. 5) and to declare that species and therefore the genus *Otarorhyncha* as *nomina dubia*. The specimen illustrated by Savage (2002, fig. 707.3) as *Otarorhyncha otarica* is the same one as shown by Nikiforova & Popov (1981, pl. 7, fig. 11) and came from the Akkol Beds of Dulankara Mountains and therefore does not belong to '*Rhynchotrema*' *otarica* in a strict sense. Savage (2006, p. 2724) also discussed the genus *Altaethyrella* Severgina, 1978, at length and reproduced Severgina's original illustrations. He concluded that *Altaethyrella* is undoubtedly a *nomen dubium* and thus cannot be used as a substitute for *Otarorhyncha* for costate ancystrorhynchids with a septiform cardinal process, a well-developed dorsal median fold and ventral sulcus. For some time we considered the generic name *Altaethyrella* as applicable to some ancystrorhynchid rhynchonellides from Kazakhstan and China. However, important comments were made by Savage (2006) in the latest edition of the 'Treatise', and we have now concluded that *Altaethyrella* is best treated as a '*nomen dubium*'. That does not mean that the taxon is not formally applicable according to ICZN rules, but it means that the taxon cannot be recognised because of inadequate information given in the original description. L. E. Popov saw the type specimens about forty years ago, and, although the shells were clearly rhynchonellide, details of their interiors were unavailable, and any new information on the types or the collection of topotypes is extremely unlikely to become available in the foreseeable future. Thus in the present paper we do not claim that the taxon name *Altaethyrella* is invalid, but following the discussion of Savage (2006), we consider that the name cannot sensibly be applied to any specimens outside the type locality in the Altai Region. This state of affairs is unlikely to change in the foreseeable future. Thus, since *Altaethyrella megala* Severgina is a *nomen dubium*, and the latter group are without a generic home, we therefore assign them here to the new genus *Lydirhyncha* with *Rhynchotrema zhejiangensis* as the type species. *Lydirhyncha* differs from other genera of the Ancistrorhynchidae (Savage, 2002), in having a septiform cardinal cardinal process, which is absent in all other genera in the family, a strong dorsal median fold, and a ventral sulcus originating in the umbonal area.

Lydirhyncha tarimensis (Sproat & Zhan, 2018)

Figures 27P, Q

1956 *Rhynchotrema otarica* sp. nov. Rukavishnikova (*pars*), p. 156, pl. 5, figs 6, 9, 10 (*non* pl. 5, figs 7, 8).

1964 *Rhynchotrema zhejiangensis* Wang *in* Wang & Jin, p. 46, pl. 12, figs 7–11.

1981 *Otarorhyncha otarica* (Rukavishnikova) Nikiforova & Popov, p. 62, pl. 7, figs 10–15, pl. 8, fig. 1, text-fig. 1a, 2 (*non* pl. 7, figs 13, 14).

1986 *Rhynchotrema otarica* Rukavishnikova; Misius, p. 202, pl. 12, figs 4–5.

2000 *Altaethyrella otarica* (Rukavishnikova, 1956) Popov, Nikitin & Cocks, p. 862, pl. 4, figs 19–23.

2006 *Altaethyrella otarica* (Rukavishnikova, 1956); Nikitin, Popov & Bassett, p. 267, figs 25.8–11, 33, 34.1–7.

2006 *Altaethyrella otarica* (Rukavishnikova, 1956); Popov & Cocks, p. 278, fig. 6C, D.

2018 *Altaethyrella tarimensis* new species Sproat & Zhan, 2018, p. 1009, figs 1, 4–7; tables 1–3.

Holotype. – NIGP 167281, articulated shell from the Hadabulaktag Formation (late Katian) of the Kuruktag region, southern Xingjiang, north-west China.

Material. – Degeres Member of the Dulankara Formation, Dulankara Mountains, Locality 546: BC 57780 (Popov & Cocks, 2006, fig. 6D), dorsal internal mould. Locality 827: articulated CNIGR 34/11774. Locality 835: dorsal external mould 33/11774. Locality 836: ventral internal mould CNIGR 31/11774, dorsal internal mould CNIGR 32/11774. Kopkurgan Formation, 10 km south-west of Alakol Lake, West Balkhash, Locality 735: BC 63881 (Fig. 27Q; Ld, 14.7; W, 18.0; Sw, 5.7), dorsal valve; BC 63872 (Fig. 27P), dorsal internal mould.

Description of the Kazakh material. – Shell dorsibiconvex, subpentagonal to subtriangular outline, 70–100% as long as wide. Ventral beak strongly incurved, dorsal umbonal area swollen. Radial ornament costate with 3 costae in ventral sulcus, 4 on dorsal median fold and 6–10 on flanks, a total of 15–24 ribs. Concentric ornament of dense, evenly spaced faint filae, often poorly preserved. Ventral valve moderately and evenly convex with strongly incurved, pointed beak. Delthyrium open, triangular. Ventral sulcus with smooth, steep lateral slopes, originating at about 3 mm from the umbo and terminating with a high, trapezoidal tongue in mature individuals. Dorsal valve strongly and evenly convex with a swollen umbonal area and a median fold with steep lateral slopes originating about 3 mm from the umbo. Ventral interior with oblique cyrtomatodont teeth supported by vestigial dental plates close to the shell wall (Nikiforova & Popov 1981, text-fig. 2; Nikitin *et al.* 2006, figs 33A, B). Ventral muscle field strongly elongate, rounded, subtriangular, surrounded anteriorly and laterally by distinct muscle bounding ridges (Nikiforova & Popov 1981, pl. 7, fig. 10). Adductor muscles narrow, subtriangular, raised anteriorly, dividing narrow, strongly impressed diductor scars of about equal length. Delthyrial cavity floor occupied by a strongly impressed pedicle collar. Dorsal valve with short, disjunct hinge plates. Cardinal process faint, septiform (Fig. 27P). Long, gently curved dorsolaterally radulifer crura with a slightly grooved outer surface. Dorsal adductor muscle field weakly impressed, bisected by a faint median ridge terminating at the mid-valve.

Remarks. – Comparison of the shells such as *Lydirhyncha otarica* (Popov *et al.*, 2000; Nikitin *et al.* 2002; Popov & Cocks 2006) from other Kazakh terranes, with *Lydirhyncha tarimensis* (Sproat & Zhan, 2018) shows that they are identical in internal morphology, including vestigial dental plates, a septiform cardinal process, and a short, disjunct hinge plate, while in the variable external morphology of the shell, including sizes, proportions and characters of radial ornament: the Kazakh and Chinese populations significantly overlap each other and are thus considered conspecific. Remarkably, the statistics in Nikitin *et al.* (2006, table 35) closely match those given for *Lydirhyncha tarimensis* by Sproat & Zhan (2018, tables 1–3) and are also well within the wide range of variation in the Kazakh populations of *Lydirhyncha* reported by Nikiforova & Popov (1981) and Popov *et al.* (2000).

A few dorsal valves from the Kopkurgan Formation also show the characteristic features of *Lydirhyncha tarimensis*. The *Lydirhyncha* shells from the Otar Beds of southern Chu-Ili Range (Nikiforova & Popov 1981; Popov *et al.* 2000; localities 131, 542, and 837) tend to be less transverse (L/W average values varying from 96% [S, 13; N, 8] to 102% [S, 10.5; N, 15]) than shells from the Degeres and Akkol Beds (L/W average values varying from 83% [S, 12; N, 11] to 86% [S, 18; N, 8]); however, that can be explained by a large proportion of juvenile shells in the Otar samples, which usually have a more elongate, subtriangular shell. Strong variations in maximum sizes and proportions of the shell, and width of the dorsal fold/ventral sulcus in relation to the shell width were reported by Nikitin *et al.* (2006, tables 34–36) for populations from the Koskarasu and Otar beds of the Boshchekul Island arc.

Lydirhyncha tarimensis (Sproat & Zhan, 2018) is closely similar to *Lydirhyncha zhejiangensis* in its shell outline and radial ornament, and the only substantial difference from topotypes is in the relatively more strongly biconvex shell, while the taxonomic significance of such characters as the slightly higher dorsal fold, more convex shell flanks are doubtful, although the average and maximum shell sizes are very variable between different Kazakh populations of *Lydirhyncha tarimensis*. The species attribution of the shells described as 'Rhynchotrema' otarica by Klenina *et al.* (1984) from the Taldyboi Formation of the Chingiz Range and by Nikiforova (1978) from the Archalyk Bedres (late Katian) of the Zerafshan Range, Uzbekistan, requires further study, but generic assignment of the Uzbek shells to *Lydirhyncha* appears undoubted.

Family Sphenotretidae Savage, 1996

Genus *Baitalorhyncha* n. gen.

Type species. – *Baitalorhyncha rectimarginata* n. sp., see below.

Derivation of name. – After the Baital peninsula on the west coast of Lake Balkhash near the type locality.

Diagnosis. – Shell subtriangular, rostrate, subequally biconvex with rectimarginate anterior commissure, ornamented with simple, rounded ribs and faint, crowded concentric filae. Delthyrium narrow, open. No trace of sulcus or median fold on either valve.

Ventral interior with thin, straight, narrowly divergent dental plates enclosing strongly impressed pedicle collar. Dorsal interior with small, divided hinge plates. Dorsal adductor muscle field elongate suboval, bounded laterally by faint ridges and bisected medially by a long faint myophragm, but no median septum.

Species questionably assigned. – *Dorytreta insignis* Nikitin & Popov *in* Klenina *et al.*, 1984, from the Bestamak Formation (Sandbian) of the Chingiz Range, Kazakhstan, which has no dorsal sulcus, anterior commissure weakly uniplicate with a poorly defined dorsal median fold and ventral sulcus in large individuals, characters which are unknown in other genera of Sphenotretidae.

Remarks. – While the morphology of the cardinalia and characters of the radial ornament suggest attribution of the new taxon to the Family Sphenotretidae, *Baitalorhyncha* differs from *Sphenotreta* Cooper, 1956, and *Dorytreta* Cooper, 1956, the only other genera in the family, in the rectimarginate anterior commissure and in the absence of a distinct sulcus and median fold on both valves.

Baitalorhyncha rectimarginata n. sp.

Plate 21, figures 1–16

Derivation of name. – In recognition of the characteristic rectimarginate anterior commissure.

Holotype. – BC 62402 (Lv, 12.0, Ld, 10.8, W, 11.4, T, 6.6; Pl. 21, figs 1–4) articulated shell, from Locality 8121, Berkutsyur Formation (early Sandbian), about 7 km south-west of Lake Alakol, West Balkhash.

Paratypes. – Locality 8234: 25 ventral and 23 dorsal valves preserved as internal and external moulds, including BC 62386 (Pl. 21, fig. 5), BC 62391 (Pl. 21, fig. 8), BC 62392 (Pl. 21, fig. 6), BC 62393 (Pl. 21, fig. 7), BC 62394, BC 63817c, BC 63831a, BC 63848a (Ld, 5.2; W, 4.1), BC 63848b (Ld, 6.4; W, 6.9), BC 63851a, b, BC 63853a (Lv, 6.0; W, 5.5; Pl. 21, fig. 15), BC 63853b (Pl. 21, fig. 10), BC 63854a (Pl. 21, fig. 13), BC 63854b (Pl. 21, fig. 14), BC 63855, BC 63857a (Ld, 4.0; W, 3.9; Pl. 21, fig. 12), BC 63857b (Pl. 21, fig. 11), BC 63858a, BC 63859b (Pl. 21, fig. 16), BC 63860 (Lv, 6.1; W, 5.9), BC 63861, BC 63862, BC 63865b, (Pl. 21, fig. 9), BC 63867b, BC 63870c, BC 63892b, NMW 98.28G.2159, ventral external mould, NMW 98.28G.2160, dorsal internal mould.

Diagnosis. – Subequally biconvex, rostrate, elongate, subtriangular, rectimarginate shell with maximum width almost at the anterior margin, costate, ornamented with 11–15 subangular ribs, no fold and sulcus.

Description. – Shell subequally biconvex, slightly elongate, subtriangular, about 103–109% as long as wide, with maximum width near the anterior margin, and slightly more than half as thick as long. Hinge line very short, curved; anterior commissure rectimarginate. Ventral valve moderately and evenly convex, with a small, pointed beak slightly curved posterodorsally. Delthyrium very small, open, triangular, with narrow deltidial plates on flanks. Dorsal valve evenly convex with a small, curved beak not extending beyond the hinge line. Radial ornament costate, with 11–15 subangular ribs originating at the umbo. Concentric ornament of faint crowded filae, about 7–8 per mm. Ventral valve with small teeth supported by thin, straight slightly divergent, extended anteriorly faint muscle bounding ridges flanking the elongate subtriangular ventral muscle field extending anteriorly almost to the mid-valve. Ventral diductor scars bisected posteriorly by a faint myophragm, then anteriorly divided by the subrectangular adductor scar, but not enclosing them (Pl. 21, fig. 8). Posterior two-thirds of the delthyrial cavity floor occupied by the pedicle collar (Pl. 21, figs 8, 13, 14). Dorsal interior with minute, divided hinge plates. Dorsal adductor muscle field elongate and suboval, bounded laterally by faint ridges evenly curved outward and bisected medially by the faint, long myophragm, extending slightly beyond the mid valve (Pl. 21, figs 5, 7, 12). Dorsal mantle canal system lemniscate with straight vascula media divergent under 115° (Pl. 21, fig. 12).

Remarks. – The only comparable species is *Dorytreta? insignis* Nikitin & Popov in Klenina *et al.*, 1984, but unlike *Baitalorhyncha rectimarginata* that has a weakly defined ventral sulcus and dorsal median fold which appear close to the anterior margin of mature individuals, as well as finer radial ornament with 14–16 ribs.

Family Oligorhynchidae Cooper, 1956

Genus *Paraoligorhyncha* Popov *in* Nikiforova & Popov, 1981

Type species (by original designation). – *Paraoligorhyncha reducta* Popov *in* Nikiforova & Popov,

1981, Dulankara Formation (Katian), Dulankara Mountains, Chu-Ili Terrane, Kazakhstan.

Paraoligorhyncha? sp.

Figure 27M, O

Material. – Baigara Formation, Locality 8121: one articulated shell; NMW 98.28G.1996.

Remarks. – This specimen is remarkable among Sandbian rhynchonellides in having a smooth, rostrate shell with a parasulcate anterior commissure and a pair of low, broad plications on its flanks, and this distinctive shell morphology suggests attribution to *Oligorhyncha* Cooper, 1956, or *Paraoligorhyncha*. Those two genera are discriminated mainly on the absence or presence of a cardinal process respectively, but they are almost indistinguishable externally. Thus, in the absence of data on the interiors of both valves, the generic assignation of the specimen is provisional and mainly based on the fact that *Oligorhyncha* is as yet unknown from the Kazakh terranes, while *Paraoligorhyncha* is endemic to Chu-Ili.

Order Atrypida Rzhonsnitskaya, 1960
Suborder Atrypidina Moore, 1952
Superfamily Atrypoidea Gill, 1871
Family Atrypidae Gill, 1871

Genus *Qilianotryma* Xu, 1979

Type species (by original designation). – *Qilianotryma mirabile* Xu, 1979, from the Koumenzi Formation (Katian) of Qilian, western Qinghai, China.

Qilianotryma cf. *Q. suspectum* (Popov *in* Nikiforova *et al.,* 1982)

Plate 10, figure 19

cf. 1982 *Euroatrypa suspecta* Popov, sp. nov., Nikiforova, Popov & Oradovskaya, p. 67, pl. 6, figs 9–12.

Remarks. – The single dorsal internal mould, NMW 98.28G.1966, from Locality 1501 in an unnamed formation near the Ergibulak Well, has well preserved cardinalia, faint dorsal median ridge and strong dorsal median fold dorsal median fold sharpened laterally by plications which leaves no doubt about its generic assignment. Although it looks indistinguishable from dorsal valves of *Qilianotryma suspectum* from the Degeres Member of the Dulankara Formation in the Dulankara Mountains, southern Chu-Ili Range (Nikiforova *et al.* 1982; Popov *et al.* 1999); in the absence of data on the ventral valve the species can only be attributed to it with a query.

Suborder Lissatrypidina Copper, 1996

Superfamily Protozygoidea Copper, 1986

Family Kellerellidae n. fam.

Diagnosis. – Shell astrophic, ventribiconvex, smooth or finely capillate, parasulcate, rarely, uniplicate or rectimarginate. Ventral valve with small open, triangular delthyrium. Ventral interior with a short, thin dental plate located close to posterolateral valve margins, cardinalia very small with disjunct outer hinge plates. Inner hinge plates rudimentary or absent, a small septalium supported by the thin median septum may be present. Spiralia with laterally directed whorls located posterodorsally, and simple, short, posteroventrally directed jugal processes.

Genera assigned. – *Kellerella* and *Nikolaispira*, both erected by Nikitin & Popov *in* Nikitin, Popov & Holmer (1996), from the Sandbian and Katian of Kazakhstan; *Costistriispira* n. gen., from the Berkutsyur Formation, West Balkhash (see below).

Remarks. – *Kellerella* and *Nikolaispira*, which make their first appearances in the late Sandbian, are among the oldest spirebearers and are moreover probably the earliest brachiopods with laterally directed spiral cones characteristic of both athyridides and spiriferides. The two genera were originally assigned to the atrypides (Family Meristellidae) mainly because of the astrophic shell and laterally directed spiralia; however, as noted by Alvarez (2006), externally they are closely comparable to other early smooth atrypides, such as *Cyclospira* or *Rozmanospira*, but have short disjunct jugal processes instead of the complex athyridid jugum, a minute cardinalia lacking crural plates, and are commonly without a septalium and supporting septum. Similar cardinalia can be seen in *Rozmanospira* (Popov *et al.* 1999), which differs mainly in the absence of a jugum and a very

simple partial spiral whorl. The shell shape, cardinalia, characters of spiralia, and jugal processes in *Kellerella* show much similarity to the earliest spiriferides, and in particular to juveniles of *Protospirifer praecursor* (Rong *et al.*, 1994, fig. 5A), which also appear to be astrophic. *Costistriispira* is apparently the oldest genus of the family and is closely similar to *Kellerella*, but differs in having finely capillate radial ornament; however, the characteristics of its spiralia are unknown.

Genus *Costistriispira* n. gen.

Derivation of name. – After the characteristic ornament of radial ribs and superimposed capillae.

Type and only species. – *Costistriispira proavia* n. sp. from the Kopkorgan Formation (early Sandbian), area 4 km south-west of Lake Alakol, West Balkhash.

Diagnosis. – Shell ventribiconvex, slightly elongate, suboval, with a short curved hinge-line and gently uniplicate anterior commissure. Ventral valve with indistinct sulcus appearing close to the anterior margin of large individuals. Radial ornament paucicostate of low, rounded ribs, increasing in number by bifurcation. Ventral interior with small teeth, short straight divergent dental plates, delthyrial cavity floor occupied by the pedicle collar. Dorsal interior with divided hinge plates. No cardinal process and median septum. Indistinctly impressed dorsal adductor scars bisected by a faint myophragm.

Remarks. – *Costistriispira* differs from other genera of the family Kellerellidae (*Kellerella* and *Nikolaispira*) in its paucicostate shell, in having dense radial capillae on the shell surface, and a poorly defined ventral sulcus and dorsal median fold. It also differs from *Nikolaispira* in the absence of a septalium and a dorsal median ridge. From the superficially similar enigmatic *Manosia* Zeng *in* Wang 1983, it differs in having capillate radial ornament superimposed on ribs originating at some distance from the umbo, a very short, curved hinge line, and in the absence of delthyrial covers.

Costistriispira proavia n. sp.

Figures 28A–P; 29J, K

Derivation of name. – After Latin, *proavia* – grandmother.

Holotype. – BC 63892a (Ld = 9.0, W = 9.6; Figs 28 M, 29J), dorsal internal mould, Locality 8234, from the Berkutsyur Fm Formation (early Sandbian), area 4 km south-west of Lake Alakol, West Balkhash.

Material. – Locality 8234, nine ventral and 13 dorsal valves, including paratypes, BC 63817a (Fig. 28 H), BC 63817b (Fig. 28D), BC 63831b (Fig. 28C), BC 63831c, BC 63831d (Lv, 7.4; W, 6.0), BC 63831e, (Ld, 5.2; W, 4.8), BC 63852 (Fig. 28F), BC 63856 (Ld, 8.5; W,9.7; Fig. 28E), BC 63858b, BC 63859a (Ld, 6.6; W,6.6; Fig. 28O), BC 63863 (Fig. 28I), BC 63864 (Fig. 28K), BC 63867a (Ld, 8.2; W, 7.5; Fig. 28B), BC 63868b (Ld, 9.7; W, 9.3; Figs 28G; 29K), BC 63869a (Fig. 28L, P), BC 63869b (Lv, 6.4; W, 5.1; Fig. 28A), BC 63870a (Fig. 28J), BC 63870b (Fig. 28N). NMW 98.28G.2161, dorsal internal mould; NMW 98.28G.2165, ventral external mould.

Diagnosis. – As for the genus.

Description. – Shell ventribiconvex, slightly elongate suboval, with a short, curved hinge-line and gently uniplicate anterior commissure. Ventral valve moderately and evenly convex, with small, posterodorsally curved beak and indistinct sulcus appearing close to the anterior margin of large individuals. Dorsal valve lateral profile gently convex, with maximum height slightly posterior to mid-length. Paucicostate radial ornament of low, rounded ribs originating at 2–3 mm from the umbo, about 12 in the posterior part of the shell, increasing by bifurcation up to 20 in the largest individuals and gradually fading close to the posterolateral corners of both valves. Faint densely-spaced radial capillae, varying from 9 to 14 per mm (Fig. 28L). Concentric ornament of a few thin growth lamellae and faint regular concentric fila about 8 per mm (Fig. 28L, P). Ventral interior with small teeth and short thin, straight, divergent dental plates (Fig. 28D, H). Delthyrial cavity floor occupied by the pedicle collar (Fig. 28D). Ventral muscles and mantle canals not impressed. Dorsal interior with a divided hinge plate and weakly impressed adductor muscle field bisected by the myophragm terminated slightly behind the mid-valve (Fig. 28B, G).

Remarks. – While paucicostate radial ornament is a general characteristic for *Costistriispira proavia*, that feature is very variable within the studied population; a few individuals are almost completely smooth (Pl. 25, figs A, C, F), while in some others ribs are only developed in the umbonal area (Pl. 25, fig. E).

Costistriispira? sp.

Plate 15, figure 18

Remarks. – These subcircular dorsal external moulds BC 62409 and BC 62410 (Pl. 15, fig. 18) from Locality 8122 in the Kopkurgan Formation (Sandbian), have paucicostate radial ornament of 17 low, rounded ribs increasing by bifurcation in the anterior part of the shell and superimposed faint regular concentric fila strongly recalling *Costistriispira*, but any radial capillae are not preserved and there is no data on their dorsal interior. While no other Sandbian taxon from Chu-Ili except *Costistriispira proavia* has comparable radial ornament, the available material is insufficient for more precise determination.

Family Cyclospiridae Schuchert, 1913

Genus *Rozmanospira* Popov, Nikitin & Sokiran, 1999

Type (by original designation) and only species. – *Oligorhynchia mica* Nikitin & Popov *in* Klenina *et al.*, 1984, see below.

Remarks. – While this genus was formerly considered to lie within the Cyclospiridae and Protozygoidea (Popov *et al.* 1999; Copper 2002), it probably represents a basal member of the lineage of the endemic Kazakh smooth atrypides assigned here to the family Kellerellidae. It is similar to the latter in its general shell morphology and cardinalia, but differs in its simple brachial supports without a jugum and with a single partial spiral whorl.

Rozmanospira mica (Nikitin & Popov *in* Klenina, Nikitin & Popov 1984)

Figure 29A–H

1984 *Oligorhynchia mica* Nikitin & Popov, sp. nov., Klenina, Nikitin & Popov, p. 156, figs 4, 5, text-figs 38, 39.

1999 *Rozmanospira mica* (Nikitin & Popov) Popov, Nikitin & Sokiran, p, 645, pl. 4, figs 1–8, text-fig. 9.

Holotype. – CNIGR 197/12095, Locality 564, articulated shell, from the Bestamak Formation (Sandbian), east side of the Chagan River opposite to Bestamak village, Chingiz Range.

Material. – Berkutsyur Formation, Locality 8121: 11 articulated shells, including NMW 98.28G.1989 (Fig. 29A–D), 1990 (Fig. 29E–H), 1991–1994.

Remarks. – *Rozmanospira mica* was originally documented from the Sandbian of the Chingiz – Tarbagatai island arc system, but it is also present in the early Sandbian of the Chu-Ili Terane where it occurs in algal limestones with abundant dasyclad algae. These smooth, ventribiconvex shells are distinct in their small size and parasulcate anterior commissure.

Acknowledgements

The material described here was collected over a period of more than two decades, often with the assistance of O. P. Kovalevskii, I. F. Nikitin, A. V. Alperovich, D. T. Tsai, and E. A. Vinogradova, who also shared with us their field notes, maps, and other valuable field documentation. Many of them have died, yet they all stay as warmly remembered former colleagues who helped to make this paper possible. L. E. Popov acknowledges curatorial and other support from the National Museum of Wales and L. R. M. Cocks from The Natural History Museum. We are also very grateful for the constructive comments on the manuscript from Yves Candela (Edinburgh, Scotland) and Jisuo Jin (London, Ontario). Financial support for the publication of this issue of Fossils and Strata was provided by the Lethaia Foundation.

References

Alvarez, F. 2006: Athyridida. *In* P.A. Selden (ed.): *Treatise on Invertebrate Paleontology. Part H Brachiopoda Revised, volume 6*, 2742–2771. The Geological Society of America, Boulder, Colorado and the University of Kansas Press, Lawrence.

Apollonov, M. K., Bandaletov, S. M. & Nikitin, I. F. 1980: *The Ordovician – Silurian Boundary in Kazakhstan.* Nauka, Alma-Ata, 300 pp. [In Russian].

Bancroft, B. B. 1928: On the notational representation of the rib-system in Orthacea. *Memoirs of the Manchester Literary and Philosophical Society 72*, 53–90, pls 1–3.

Bassett, M. G., Ghobadi Pour, M., Popov, L. E. & Kebria-ee Zadeh, M. 2013: First report of craniide brachiopods in the Palaeozoic of Iran (*Pseudocrania*, Ordovician), and Early to Mid-Ordovician biogeography of the Craniida. *Palaeontology 56*, 209–216.

Bassett, M. G., Popov, L. E. & Holmer, L. E. 2002: Brachiopods: Cambrian – Tremadoc precursors to Ordovician radiation events. *In* Crame, J. A. & Owen, A. W. (eds): *Palaeobiogeography and Biodiversity Change: A Comparison of the Ordovician and Mesozoic–Cenozoic Radiations.* The Geological Society, London, Special Publications 194, 13–23.

Bassler, R. S. 1919: Report on the Cambrian and Ordovician formations of Maryland. *Maryland Geological Survey, Special Publication for 1919*, 1–424.

Bazhenov, M. L., Collins, A. Q., Degtyarev, K. E., Levashova, N. M., Mikolaichuk, A. V., Pavlov, V. E. & Van der Voo, R. 2003: Paleozoic northward drift of the North Tian Shan (Central Asia) as revealed by Ordovician and Carboniferous paleomagnetism. *Tectonophysics 366*, 113–141.

Bazhenov, M. L., Levashova, N. M., Degtyarev, K. E, Van der Voo, R., Abrajevitch, A. V. & McCausland, P. J. A. 2012: Unraveling the early–middle Paleozoic paleogeography of Kazakhstan on the basis of Ordovician and Devonian paleomagnetic results. *Gondwana Research 22*, 974–991.

Billings, E. 1858: Black River Fauna: descriptions of Devonian and Ordovician fossils. *Geological Survey of Canada Report of Progress*, [for 1857], 147–192.

Borissiak, M. A. 1955: Stratigraphy and fauna of Ordovician and Silurian deposits from the central part of Kazakhstan. *Materialy Vesesouznogo nauchno-issledovatelskiogo geologicheskogo instituta 56*, 1–107, 13 pls. [in Russian].

Borissiak, M. A. 1956: The genus *Kassinella*. *Materiali vsesoyuznova nauchna-issledovatelski Geologicheskova Instituta, Moskva, (new series) 12*, 50–52. [In Russian].

Borissiak, M. A. 1972: New Late Ordovician strophomenides from Eastern Kazakhstan. *In* Zenina, V. (ed.): *Novyie vidy drevnikh rastenii i bespozvonochnykh SSSR*, 182–184. Nauka, Moscow. [In Russian].

Brongniart, A. 1828: *Prodrome d'une historie des vegetaux fossiles.* F. G. Levrault, Paris. 223 pp.

Boucot, A. J. 1975: *Evolution and extinction rate controls.* Elsevier, Amsterdam, xv + 427 pp.

Clarke, J. M. 1902: A new genus of Paleozoic brachiopods, *Eunoa* with some considerations there on *Discinocaris*, *Spathiocaris* and *Cardiocaris*. *New York State Museum Bulletin 52*, 606–615.

Cocks, L. R. M. 1968: Some strophomenacean brachiopods from the British Lower Silurian. *Bulletin of the British Museum (Natural History), Geology 15*, 283–324, pls 1–14.

Cocks, L. R. M. 1970: Silurian brachiopods of the Superfamily Plectambonitacea. *Bulletin of the British Museum (Natural History), Geology 19*, 139–203, pls 1–17.

Cocks, L. R. M. 2005: Strophomenate brachiopods from the late Ordovician Boda Limestone of Sweden: their systematics and implications for palaeogeography. *Journal of Systematic Palaeontology 3*, 243–282, pls 1–12.

Cocks, L. R. M. 2008: A revised review of Britsh Lower Palaeozoic brachiopods. *Monograph of the Palaeontographical Society, London 161*, 1–276, pls 1–10.

Cocks, L. R. M. 2013: Generic identities and relationships within the brachiopod family Sowerbyellidae. *Palaeontology 56*, 167–181.

Cocks, L. R. M. 2019: Llandovery brachiopods from England and Wales. *Monograph of the Palaeontographical Society, London 172*, 1–262, pls 1–41.

Cocks, L. R. M. & Rong, J. 1989: Classification and revision of the brachiopod superfamily Plectambonitacea. *Bulletin of the British Museum (Natural History), Geology 45*, 77–163.

Cocks, L. R. M. & Rong, J. 2000: Order Strophomenida. *In* Kaesler, R. L. (ed.): *Treatise on invertebrate paleontology. Part H Brachiopoda Revised*, volume 2, 216–348. Geological Society of America, Boulder, Colorado and the Kansas University Press, Lawrence.

Colmenar, J. & Rasmussen, C. M. Ø. 2018: A Gondwanan perspective on the Ordovician Radiation constrains its temporal duration and suggests first wave of speciation, fuelled by Cambrian clades. *Lethaia 51*, 286–295.

Cooper, G. A. 1930: New species from the Upper Ordovician of Percé. *American Journal of Science 20*, 48–53, pl. 1.

Cooper, G. A. 1956: Chazyan and related brachiopods. *Smithsonian Miscellaneous Collections 127*, 1–1245, pls 1–260.

Copper, P. 2002. Atrypida. *In* Kaesler, R. L. (ed.): *Treatise on Invertebrate Paleontology, Part H, Brachiopoda, Revised 4*, 1377–1474. Geological Survey of America, Boulder, Colorado and the University of Kansas Press, Lawrence.

Dalman, J. W. 1828: Uppställning och Beskrifning af de i Sverige funne Terebratuliter. *Konglinga Svenska Vetenskapsakademiens Handlingar [for Ar 1827] 3*, 85–155, pls 1–6.

Davidson, T. 1853: British fossil Brachiopoda, volume I. Introduction. *Monograph of the Palaeontographical Society 21*, 136 pp., pl. 9.

Davidson, T. 1883: A monograph of the British Brachiopoda, Volume V, Part 2, Silurian supplement. *Monograph of the Palaeontographical Society 27*, 135–342, pls 8–17.

Degtyarev, K. E. & Ryazantsev, A. V. 2007: Cambrian arc-continent collision in the Paleozoides of Kazakhstan. *Geotectonics 41*, 63–86. [In Russian].

Dubinina, S. V., Orlova, A. P. & Kurkovskaya, L.A. 1996: Co-occurrence of the conodonts and graptolites in the Lower Ordovician siliceous-terrigenous deposits of northern Betpak-Dala (Kazakhstan). *Bulletin of Moscow Society of Naturalists 71*, 44–50. [In Russian].

Eichwald, C. E. 1840: *Ueber das Silurische Schichtensystem in Estland.* Medizishen Akademie, St. Petersburg, 210 pp.

Emig, C. C. 1997: Ecology of inarticulated brachiopods. *In* Kaesler, R. L. (ed.): *Treatise on invertebrate paleontology, Part H Brachiopoda* Revised 1, 473–495. Geological Society of America, Boulder, Colorado and Kansas University Press. Lawrence.

Emmons, E. 1842: *Geology of New York. Part II, comprising the survey of the second geological district.* White & Vischer, Albany, 437 pp.

Esenov, Sh. E., Galitskii, V. V., Kostenko, N. N. & Shlygin, A. E. (eds) 1971: *Southern Kazakhstan. Geology of USSR, 40 Part 1*, 409pp., *Part 2*, 286 pp. [In Russian].

Foerste, A. F. 1909: Fossils from the Silurian formations of Tennessee, Indiana and Kentucky. *Bulletin of the Denison University Science Laboratories 14*, 61–116, pls 1–4.

Foerste, A. F. 1912: *Strophomena* and other fossils from Cincinnatian and Mohawkian horizons, chiefly in Ohio, Indiana and Kentucky. *Bulletin of the Denison University Science Laboratories 17*, 17–173, pls 1–8.

Foerste, A. F. 1914: Notes on the Lorraine faunas from New York and the Province of Quebec. *Bulletin of the Denison University Science Laboratories 17*, 247–328, pls 1–5.

Fortey, R. A. & Cocks, L. R. M. 2003: Palaeontological evidence bearing on global Ordovician–Silurian continental reconstructions. *Earth-Science Reviews 61*, 245–307.

Franeck, F. & Liow, L. H. 2019: Dissecting the paleocontinental and paleoenvironmental dynamics of the great Ordovician biodiversification. *Paleobiology 45*, 221–234.

Fu, L. 1982: Brachiopoda. *In* Xian Institute of Geology and Mineral Resources (eds): *Palaeontological Atlas of North-West China, Shaanxi- Ganxu-Ningxia Volume: Part 1, Precambrian and Early Paleozoic*, 95–178, pls 30–45. Geological Publishing House, Beijing. [In Chinese].

Ghobadi Bassett, M., Popov, L. E., McCobb, L. M. E. & Percival, I. G. 2011: New data on the Late Ordovician trilobite faunas of Kazakhstan: implications for biogeography of tropical peri-Gondwana. *In* Gutiérrez-Marco, J. C., Rabano, I. & García-Bellido, D. (eds): *Ordovician of the World. Cuadernos del Museo Geominero 14*, 171–177. Instituto Geológico y Minero de España, Madrid.

Ghobadi-Pour, M., Bassett, M. G., Popov, L. E. & Vinogradova, E. V. 2009: Middle Ordovician (late Darriwilian) trilobites from the northern Betpak-Dala Desert, central Kazakhstan. *Memoirs of the Association of Australasian Palaeontologists 37*, 327–349.

Gill, T. 1871: Arrangement of the families of molluscs prepared for the Smithsonian Institution. *Smithsonian Miscellaneous Collections 227*, 1–49.

Gridina, N. M., Magretova, L. I. & Evseenko, R. D. 2004: On the stratigraphical subdivision of the type section of the lower to middle Ordovician Sarybidaik Formation (north-eastern Central Kazakhstan). *Geologiya i okhrana nedr, KazGEO 2004*, 22–26. [In Russian.]

Hall, J. 1847: Description of the organic remains of the lower division of the New-York System. *Contributions to the Palaeontology of New York, New York Geological Survey, Palaeontology of New York 1*, 1–338, pls 1–97.

Hall, J. 1859: Observations on genera of Brachiopoda. Contributions to the Palaeontology of New York, *New York State Cabinet of Natural History, 12th Annual Report*, Albany, 8–110.

Hall, J. 1883: Brachiopoda. Plates and explanations. *New York State Geologist, 2nd Annual Report for 1882.* Albany, New York, 17pp, 61 pls.

Hall, J. & Clarke, J. M. 1892: An introduction to the study of the genera of Palaeozoic Brachiopoda. *New York State Geological Survey Palaeontology of New York 8*, 1–367, pls 1–41.

Hammer, Ø. & Harper, D. A. T. 2006: *Paleontological Data Analysis*, 351 pp. Blackwell Publishing, Malden, Oxford, Victoria.

Harper, D. A. T., Rasmussen, C. M. Ø., Liljeroth, M., Blodgett, R. B., Candela, Y., Jin, J., Percival, I. G., Rong, J. -Y., Villas, E. & Zhan, R. B. 2013: Biodiversity, biogeography and phylogeography of Ordovician rhynchonelliform brachiopods. *In* Harper, D. A. T. & Servais, T. (eds): *Early Palaeozoic biogeography and palaeogeography*. Geological Society, London, Memoir 38, 121–138.

Havlíček, V. 1981: Upper Ordovician brachiopods from the Montagne Noire. *Paleontographica, Abt. A 176*, 1–34.

Havlíček, V. 1982: Lingulacea, Paterinacea, and Siphonotretacea (Brachiopoda) in the Lower Ordovician sequence of Bohemia. *Sbornik geologickych věd, Paleontologie 25*, 9–82.

Holl, H. B. 1865: On the geological structure of the Malvern Hills and adjacent districts. *Quarterly Journal of the Geological Society, London 21*, 72–102.

Holmer, L. E., Popov, L. E., Streng, M. & Miller, J. F. 2005: Lower Ordovician (Tremadocian) lingulate brachiopods from the House and Fillmore formations, Ibex area, western Utah, USA. *Journal of Paleontology 79*, 884–906.

International Commission of Zoological Nomenclature 1999: *International Code of Zoological Nomenclature. Fourth Edition.* The International Trust for Zoological Nomenclature, London, 232 pp.

Jin, J. S. & Zhan, R. B. 2008: *Late Ordovician orthide and billingsellide brachiopods from Anticosti Island, Eastern Canada: Diversity Change through Mass Extinction.* NRC Research Press, Ottawa, 151 pp.

Jin, J. S., Zhan, R. B., Copper, P. & Caldwell, W. G. E. 2007: Epipunctae and phosphatized setae in Late Ordovician plaesiomyid brachiopods from Anticosti Island, Eastern Canada. *Journal of Paleontology 81*, 666–683.

Jones, O. T. 1928: *Plectambonites* and some allied genera. *Memoirs of the Geological Survey of Great Britain, Palaeontology 1*, 367–527, pls 21–25.

Kaesler, R. L. (ed.) 1997–2004: *Treatise on invertebrate paleontology. Part H, Brachiopoda (revised).* Geological Society of America, Boulder and the University of Kansas Press, Lawrence. 5 vols.

Keller, B. M. & Lisogor, K. A. 1954: Karakan Horizon of the Ordovician. Ordovician of Kazakhstan 1. *Trudy Instituta geologicheskikh nauk Akademii Nauk SSSR 154*, 48–98, 5 pls. [In Russian].

King, W. 1846: Remarks on certain genera belonging to the class Palliobranchiata. *Annals and Magazine of Natural History (Series 1) 18*, 26–42, 83–94.

Klenina, L. N., Nikitin, I. F. & Popov, L. E. 1984: *Brachiopods and biostratigraphy of the Middle and Upper Ordovician of the Chingiz Hills.* Nauka, Alma-Ata, 196 pp. [In Russian].

Kröger, B. & Lintulaakso, K. 2017: RNames, a stratigraphical database designed for the statistical analysis of fossil occurrences – the Ordovician diversification as a case study. *Palaeontologica Electronica 20.1.1T*, 1–12.

Kulkov, N. P. & Severgina, L. G. 1989: Stratigrafiya i brachiopody ordovika i nizhnego silura Gornogo Altaya. [Stratigraphy and brachiopods of the Ordovician and Lower Silurian of the Gorny Altai.] *Trudy Instituta geologii i geofiziki Sibirskogo otdelenija Akademii Nauk SSSR 717*, 1–223. [In Russian].

Laurie, J. R. 1991: Articulate brachiopods from the Ordovician and Lower Silurian of Tasmania. *Memoirs of the Association of Australasian Palaeontologists 11*, 1–106.

Liu, D, Xu, H. & Liang, W. 1983: Brachiopoda. *In* Nanjing Institute of Geology and Mineral Resources (eds): *Paleontological Atlas of East China, Early Paleozoic*, vol. 1, 254–286, pls 88, 89. Geological Publishing House, Beijing. [In Chinese].

Maletz, J. 2010: *Xiphograptus* and the evolution of virgella-bearing graptoloids. *Palaeontology 53*, 415–439.

M'Coy, F. 1846: *A synopsis of the Silurian fossils of Ireland collected from the several districts by Richard Griffith F.G.S.* Privately published, Dublin. 72 pp, 5 pls.

M'Coy, F. 1851: On some new Cambro-Silurian fossils. *Annals and Magazine of Natural History (Series 2) 8*, 387–409.

Mergl, M. 2002: Linguliformean and craniiformean brachiopods of the Ordovician (Trěnice to Dobrotivá formations) of the

Barrandian, Bohemia. *Acta Musei Nationalis Praguae, series B, Natural History 58,* 1–82.

Misius, P. P. 1986: *Brachiopods of the middle Upper Ordovician of North Kirgizia.* Ilim, Frunze. 254 pp. [In Russian].

Misius, P. P. & Ushatinskaya, G. T. 1977: New Ordovician and Silurian strophomenids from Kazakhstan and North Kirgiz. *Novye vidy drevnikh rastenii i bespozvonochnykh SSR 4,* 113–116. [In Russian].

Nazarov, B. B. & Popov, L. E. 1980: Stratigraphy and fauna of the siliceous-carbonate deposits of the Ordovician of Kazakhstan. *Trudy Geologicheskogo Instituta Akademii Nauk SSSR 331,* 1–190. [In Russian].

Nikiforova, O. I. 1978: Brachiopods of the Chashmankolon, Archalyk and Minkuchar beds. *Akademiia Nauk SSSR, Sibirskoe Otdelenie, Institut Geologii i Geofiziki, Trudy 397,* 102–126, pls 18–23. [In Russian].

Nikiforova, O. I. & Andreeva, O. N. 1961: Biostratigraphy of the Palaeozoic of the Siberian Platform, vol. 1. Ordovician and Silurian stratigraphy of the Siberian Platform and its paleontological basis (Brachiopoda). *Vsesoiuznyi Nauchno-Issledovatel'skii Geologicheskii Institut (VSEGEI), Trudy 56,* 1–412, 56 pl. [In Russian].

Nikiforova, O. I., Oradovskaya, M. M. & Popov, L. E. 1982: New Ordovician atrypids (Brachiopoda) from the northeastern USSR, the Taimyr Peninsula, and Kazakhstan. *Paleontologicheskii Zhurnal 1982,* 62–69. [In Russian].

Nikiforova, O. I. & Popov, L. E. 1981: New data on the Ordovician rhynchonellides of the Kazakh SSR, USSR and Central Asia. *Paleontologicheskii Zhurnal 1981,* 54–67.

Nikiforova, O. I. & Sapelnikov, V. P. 1973: Some fossil pentamerids from the Zeravshan range. *In* Sapelnikov, V. P. & Chuvashov, B. I. (eds): *Trudy Uralskii Nauchnii Tsentral Institut Geologii 99,* 64–81, pls 1–6. [In Russian].

Nikitin, I. F. 1972: *Ordovician of Kazakhstan, Part I, Stratigraphy.* Nauka, Alma-Ata, 242 pp. [In Russian].

Nikitin, I. F. 1973: *Ordovician of Kazakhstan, Part 2, Palaeogeography. Palaeotectonics.* Nauka, Alma-Ata, 100 pp. [In Russian].

Nikitin, I. F. 1974: New Middle Ordovician Plectambonitacea (Brachiopoda) from Kazakhstan. *Paleontologicheski Zhurnal 1974,* 55, 56, pls 1, 2. [In Russian].

Nikitin, I. F. (ed.) 1991: *Decisions of the Third Stratigraphical Conference on the Precambrian and Phanerozoic of Kazakhstan, Almaty, 1986. Part I. Precambrian and Palaeozoic.* Alma-Ata, 150pp. [In Russian].

Nikitin, I. F., Apollonov, M. K., Tsai, D. T. & Rukavishnikova, T. B. 1980: Ordovician system. *In* Abdulin, A. A. (ed.): *Chu-Iliiskii rudnyi poyas. Part 1. Geologiya Chu-Iliiskogo regiona,* 44–78. Nauka, Alma-Ata. [In Russian].

Nikitin, I. F., Gnilovskaya, M. B., Zhuravleva, I. T., Luchinina, V. A. & Myagkova, E. I. 1974: Anderken bioherm range and history of its origin. *Trudy Instituta Geologii I Geofyziki (IGIG), Akademii Nauk SSSR, Sibirskoe Otlodenie 84,* 122–159. [In Russian].

Nikitin, I. F. & Popov, L. E. 1983: Middle Ordovician orthides and plectambonitaceans from the northern Pri-Ishim and the Akzhar River Basin in central Kazakhstan (brachiopods). *Ezhegodnik vsesoyuznovo Paleontologicheskovo Obshchestva 28,* 34–39, pls 1–3. [In Russian].

Nikitin, I. F. & Popov, L. E. 1985: Ordovician strophomenids (Brachiopods) from north Priishim (central Kazakhstan). *Ezhegodnik vsesoyuznovo Paleonteologicheskovo Obshchestva 36,* 228–247, pls 1–3. [In Russian].

Nikitin, I. F. & Popov, L. E. 1996: Strophomenid and triplesiid brachiopods from an upper Ordovician carbonate mound in central Kazakhstan. *Alcheringa 20,* 1–20.

Nikitin, I. F., Popov, L. E. & Bassett, M. G. 2003: Late Ordovician brachiopods from the Selety river basin, north Central Kazakhstan. *Acta Palaeontologica Polonica 48,* 39–54.

Nikitin, I. F., Popov, L. E. & Bassett, M. G. 2006: Late Ordovician rhynchonelliform brachiopods of north-eastern Central Kazakhstan. *National Museum of Wales Geological Series 25,* 223–294.

Nikitin, I. F., Popov, L. E. & Holmer, L. E. 1996: Late Ordovician brachiopod assemblage of Hiberno-Salairian type from Central Kazakhstan. *GFF 118,* 83–96.

Nikitina, O. I. 1985: A new brachiopod association from the Middle Ordovician of northern Kazakhstan. *Paleontologicheski. Zhurnal 1985,* 21–29. [In Russian].

Nikitina, O. I. 1989: *Brakhiopody i biostratigrafiya srednego ordovika (llanvirna) Kazakhstana. [Brachiopods and biostratigraphy of the Middle Ordovician (Llanvirn) of Kazakhstan.]* Avtoreferat dissertatsyi na soiskaniye uchenoy stepeni kandidata geologo-mineralogicheskikh nauk. Institut Geologii i geokhimii, Sverdlovsk, 21 pp. [In Russian].

Nikitina, O. I., Popov, L. E., Neuman, R. B., Bassett, M. G. & Holmer, L. E. 2006: Mid Ordovician (Darriwilian) brachiopods of South Kazakhstan. *National Museum of Wales Geological Series 25,* 145–222.

Nikitina, O. I., Tolmacheva, T. Yu. & Ryazantsev, A. V. 2008: Stratigraphy, subdivision and main types of the Ordovician palaeobasins of northern Betpak-Dala (Central Kazakhstan). *News of National Academy of Sciences of the Republic of Kazakhstan. Series of Geology 2008,* 8–23. [In Russian].

Oehlert, D. P. 1888: Appendice sur les brachiopods. *In* Fischer, P. (ed.): *Manuel de Conchologie et de Paléontologie conchyliogique 4,* 1189–1334, pl. 15. Savy, Paris.

Öpik, A. 1933: Über Plectamboniten. *Universitatis Tartuensis (Dorpatensis) Acta et Commentationes Serie A 24,* 1–79, pls 1–12.

Pander, C. H. 1830: *Beiträge zur Geognosie des Russichen Reiches.* K. Kray, St Petersburg. 165 pp, 31 pls.

Patzkowsky, M. E. & Holland, S. M. 1997: Patterns of turnover in Middle and Upper Ordovician brachiopods of the eastern United States: a test of coordinated stasis. *Paleobiology 23,* 420–443.

Percival, I. G. 1979: Ordovician plectambonitacean brachiopods from New South Wales. *Alcheringa 3,* 91–116.

Percival, I. G. 1991: Late Ordovician articulate brachiopods from central New South Wales. *Memoir of the Association of Australasian Palaeontologists 11,* 157–178.

Percival, I. G. 2009: Late Ordovician articulate brachiopods from New South Wales. *Proceedings of the Linnean Society of New South Wales 130,* 107–177.

Phillips, J. & Salter, J. W. 1848: Palaeontological Appendix to Professor John Phillips' Memoir on the Malvern Hills. *Memoirs of the Geological Survey of Great Britain 1,* 331–396, pls 4–30.

Popov, L. E. 1977: New species of Middle Ordovician inarticulate brachiopods of Chingiz Range. *Novye vidy drevnikh rastenii i bespozvonochnykh SSSR 4,* 99–102. [In Russian].

Popov, L. E. 1980a: New species of Middle Ordovician inarticulate brachiopods of Chingiz Range. *Novyye vidy drevnikh rasteniij i bespozvonochnykh SSSR 5,* 54–57. [In Russian].

Popov, L. E. 1980b: New brachiopod species from the Middle Ordovician of the Chu-Ili Hills. *Ezhegodnik vsesoyuznovo Paleontologicheskovo Obshchestva 23,* 139–158. [In Russian].

Popov, L. E. 1985: Brachiopods of the Anderken Horizon of the Chu-Ili Hills (Kazakhstan). *Ezhegodnik vsesoyuznovo Paleontologicheskovo Obshchestva 28,* 50–68, pls 1–3. [In Russian].

Popov, L. E., Bassett, M. G., Zhemchuzhnikov, V. G., Holmer, L.E. & Klishevich, I. A. 2009: Gondwanan faunal signatures from Early Palaeozoic terranes of Kazakhstan and Central Asia: evidence and tectonic implications. *Geological Society, London, Special Publications 325,* 23–64.

Popov, L. E. & Cocks, L. R. M. 2006: Late Ordovician brachiopods from the Dulankara Formation of the Chu-Ili Range, Kazakhstan: their systematics, palaeoecology and palaeobiogeography. *Palaeontology 49,* 247–283, pls 1–6.

Popov, L. E. & Cocks, L. R. M. 2014: Late Ordovician brachiopods from the Chingiz Terrane, Kazakhstan, and their palaeogeography. *Journal of Systematic Palaeontology 12,* 687–758, pls 1–6.

Popov, L. E. & Cocks, L. R. M. 2017: Late Ordovician palaeogeography and the positions of the Kazakh terranes through analysis of their brachiopod faunas. *Acta Geologica Polonica 67,* 323–380.

Popov, L. E., Cocks, L. R. M. & Nikitin, I. F. 2002: Upper Ordovician brachiopods from the Anderken Formation, Kazakhstan: their ecology and systematics. *Bulletin of the*

British Museum (Natural History), London (Geology) 58, 13–79, pls 1–14.

Popov, L. E., Ebbestad, J. O. R., Mambetov, A. & Apayarov, F. K. 2007: A low diversity shallow water lingulid brachiopod–gastropod association from the Upper Ordovician of Kyrgyz Range. *Acta Palaeontologica Polonica 52,* 27–40.

Popov, L. E., Nikitin, I. F. & Cocks, L. R. M. 2000: Late Ordovician brachiopods from the Otar Member of the Chu-Ili Range, south Kazakhstan. *Palaeontology 43,* 833–870, pls 1–6.

Popov, L. E., Nikitin, I.F. & Sokiran, E.V. 1999: The earliest atrypides and athyrides (Brachiopoda) from the Ordovician of Kazakhstan. *Palaeontology 42,* 625–661.

Popov, L. E., Vinn, O. & Nikitina, O. I. 2001: Brachiopods of the redefined family Tritoechiidae from the Ordovician of Kazakhstan and South Urals. *Geobios 34,* 131–155.

Raymond, P. E. 1911: The Brachiopoda and Ostracoda of the Chazyan. *Annals of the Carnegie Museum 7,* 21–259.

Raymond, P. E. 1928: The brachiopods of the Lenoir and Athens formations of Tennessee and Virginia. *Bulletin of the Museum of Comparative Zoology, Harvard 68,* 293–309, pls 1–3.

Reed, F. R. C. 1905: Sedgwick Museum Notes. IV. New fossils from the Haverfordwest district. *Geological Magazine,* Decade V *2,* 97–104; 444–454; 492–501, pls 4, 23–24.

Reed, F. R. C. 1917: The Ordovician and Silurian Brachiopoda from the Girvan District. *Transactions of the Royal Society of Edinburgh 51,* 795–998, pls 1–24.

Rong, J. & Cocks, L. R. M. 1994: True *Strophomena* and a revision of the classification of strophomenind and 'stropheodontid' brachiopods. *Palaeontology 37,* 651–694, pls 1–7.

Rong, J., Fan, J., Miller, A.I. & Li, G. 2007: Dynamic patterns of latest Proterozoic-Palaeozoic-early Mesozoic marine biodiversity in South China. *Geological Journal 42,* 431–454.

Rong, J., Zhan, R. & Han, N. 1994: The oldest known *Eospirifer* (Brachiopoda) in the Changwu Formation (Late Ordovician) of western Zhejiang, East China, with a review of the earliest spiriferoids. *Journal of Paleontology 68,* 763–776.

Rong, J., Zhan, R., Huang, B., Xu, H., Fu, L. & Li, R. 2017: Ordovician brachiopod genera on type species of China. *In* Rong, J., Jin, Y., Shen, S. & Zhan, R. (eds): *Phanerozoic brachiopod genera of China,* Vol. *1,* 87–243. Science Press, Beijing, 557 pp.

Rouault, M. 1850: Note preliminaire sur une nouvelle formation decouverte dans le terrain silurien inferieur de la Bretagne. *Bulletin de la Société géologique de France 7,* 724–744.

Rowell, A. 1962: The genera of the brachiopod superfamilies Obolellacea and Siphonotretacea. *Journal of Paleontology 36,* 136–152, pls 1, 2.

Rukavishnikova, T. B. 1956: Ordovician brachiopods of the Chu-Ili Range. *Trudy Geologicheskogo Instituta, Akademii SSSR 1,* 105–168, pls 1–5. [In Russian].

Sapelnikov, V. P. 1985: *Morphological and taxonomical evolution of brachiopods (Order Pentamerida).* The Urals Scientific Center, Academy of Science of the USSR, Sverdlovsk, 188 pp., 40 pl. [In Russian].

Sapelnikov, V. P. & Rukavishnikova, T. B. 1973: Two new genera of early Pentameracea (Brachiopoda) from southern Kazakhstan. *Paleontologicheskii Zhurnal 1973,* 32–38, pl. 1. [In Russian].

Sapelnikov, V. P. & Rukavishnikova, T. B. 1975: *The Upper Ordovician, Silurian, and Lower Devonian pentamerids of Kazakhstan.* Nauka, Sverdlovsk, 227 pp., 43 pls. [In Russian].

Savage, N. M. 2002: Ancistorhynchoidea, Rhynchotrematoidea, Uncinuloidea, Camarotoechioidea. *In* Kaesler, R. L. (ed.): *Treatise on invertebrate paleontology, Part H Brachiopoda Revised 4,* 1041–1164. Geological Society of America, Boulder, Colorado and the Kansas University Press, Lawrence.

Savage, N. M. 2006: Rhynchonellida (part). *In* Selden, P. A. (ed.): *Treatise on invertebrate paleontology, Part H Brachiopoda* Revised 6, 2703–2716. Geological Society of America, Boulder and the Kansas University Press, Lawrence.

Schuchert, C. & Cooper, G. A. 1931: Synopsis of the brachiopod genera of the suborders Orthoidea and Pentameroidea, with notes on the Telotremata. *American Journal of Science (Series 5) 22,* 241–255.

Servais, T. & Harper, D. A. T. 2018: The Great Ordovician Biodiversification Event (GOBE): definition, concept and duration. *Lethaia 51,* 151–164.

Severgina, L. G. 1960: Brachiopods. *In* Khalfin, L. L. (ed.): *Biostratigraphy of the Palaeozoic of the Sayano-Altai Mountains.* Lower Palaeozoic, 400–409. *Trudy Sibirskogo nauchno-issledovatel'skogo instituta geologii, geofiziki i mineral'nogo syr'ya (SNIIGIMS), 19.* [In Russian].

Severgina, L. G. 1978: Brachiopods and stratigraphy of the Upper Ordovician of mountainous Altai region, Salair and mountainous Shoriia. *Akademia Nauk SSSR, Sibirskoe Otdelenie, Insitut Geologii i Geofiziki (IGIG), Trudy 405,* 3–41. [In Russian].

Sharpe, D. 1848: On *Trematis,* a new genus belonging to the family of brachiopodous Mollusca. *Quarterly Journal of the Geological Society, London 4,* 66–69.

Sinclair, G. W. 1945: Some Ordovician lingulid brachiopods. *Transactions of the Royal Society of Canada 39,* 55–82.

Sokolskaya, A. N. 1960: The Order Strophomenida. *In* Sarycheva, T. G (ed.): *Osnovy Paleontologii. Mshanki, Brachiopody,* 206–220. Izdatel'stvo Akademii Nauk SSSR, Moscow. [In Russian].

Sowerby, J. de C. 1839: Shells. *In* Murchison, R. I. (ed.): *The Silurian System.* 579–712, pls 1–27. John Murray, London. 768 pp., 37 pls.

Spjeldnæs, N. 1957: The Middle Ordoviacian of the Oslo Region, Norway 8. Brachiopods of the Suborder Strophomenida. *Norsk Geologisk Tidsskrift 37,* 1–214, pls 1–14.

Sproat, C. D. & Jin, J. S. 2013: Evolution of the Late Ordovician plaesiomyid brachiopod lineage in Laurentia. *Canadian Journal of Earth Sciences 50,* 872–894.

Sproat, C. D. & Jin, J. 2016: The Middle–Late Ordovician brachiopod *Plectorthis* from North America and its paleobiogeographic significance. *Palaeoworld 25,* 647–661.

Sproat, C.D. & Zhan, R. 2018: *Altaethyrella* (Brachiopoda) from the Late Ordovician of the Tarim Basin, Northwest China, and its significance. *Journal of Paleontology 92,* 1005–1017.

Stigall, A. L., Bauer, J. E., Lam, A. R. & Wright, D. F. 2017: Biotic immigration events, speciation, and the accumulation of biodiversity in the fossil record. *Global and Planetary Change 148,* 242–257.

Stubblefield, C. J. & Bulman, O. M. B. 1927: The Shineton Shales of the Wrekin district: with notes on their development in other parts of Shropshire and Herefordshire. *Quarterly Journal of the Geological Society of London 83,* 96–146.

Sutton, M. D., Bassett, M. G. & Cherns, L. 1999: Lingulate brachiopods from the Lower Ordovician of the Anglo-Welsh Basin. Part 1. *Monograph of the Palaeontographical Society, London 147,* 1–60, pls 1–8.

Tenjakova, R. G. 1989: New species of inarticulate brachiopods from the Middle Ordovician of Central Kazakhstan. *Paleontologicheskii Zhurna 1989,* 15–24. [In Russian].

Tolmacheva, T. Y. 2014: Ordovician conodont biostratigraphy and biogeography of the western part of the Central Asian Orogenic Belt. *Trudy VSEGEI, Novaya Seriya 356,* 1–263.

Tolmacheva, T. Y., Holmer, L. E., Popov, L. E. & Gogin, I. Y. 2004: Conodont biostratigraphy and faunal assemblages in radiolarian ribbon-banded cherts of the Burubaital Formation, West Balkhash Region, Kazakhstan. *Geological Magazine 141,* 699–715.

Torsvik, T. H. & Cocks, L. R. M. 2017: *Earth History and Palaeogeography.* Cambridge University Press, Cambridge. 317 pp.

Tsai, D. T. 1974: *The Early Ordovician graptolites of Kazakhstan.* Nauka Publishers, Moscow. 127 pp. [In Russian].

Tsai, D. T. 1976: *Graptolites of the Middle Ordovician of Kazakhstan.* Nauka Publishers, Alma-Ata, 76 pp. [In Russian].

Ulrich, E. O. & Cooper, G. A. 1936: New genera and species of Ozarkian and Canadian brachiopods. *Journal of Paleontology 10,* 616–631.

Ulrich, E. O. & Cooper, G. A. 1942: New genera of Ordovican brachiopods. *Journal of Paleontology 16,* 620–626, pl. 90.

Vinogradova, E. A. 2008: Granitoids of northern part of Shu-Ili Mountains and south-west Balkhash Region. *Mining Magazine 15–16,* 36–52. [In Russian].

Walcott, C. D. 1902: Cambrian Brachiopoda: *Acrotreta*; *Linnarssonnella*; *Obolus*; with descriptions of new species. *United States National Museum, Proceedings 25*, 577–612.

Wang, Y. & Jin, Y. 1964: Brachiopods. *In A Handbook of Standard Fossils from South China*. 58–78. Science Press, Beijing. [In Chinese].

Wang, Z. 1983: Brachiopoda. *In Palaeontological Atlas of North-West China, Xinjiang Province, part 2, Upper Palaeozoic*, 262–386, pl. 86–145. Science Press, Beijing[In Chinese].

Webby, B. D., Paris, F., Droser, M. L. & Percival, I. G. (eds) 2004: *The Great Ordovician Biodiversification Event*. Columbia University Press, New York. 484 pp.

Wilhem, C., Windley, B. F. & Stampfli, G. M. 2012: The Altaids of Central Asia: a tectonic and evolutionary innovative review. *Earth-Science Reviews 113*, 303–341.

Willard, B. 1928: The brachiopods of the Ottossee and Holston formations of Tennessee and Virginia. *Bulletin of the Harvard Museum of Comparative Zoology 68*, 255–292, pls 1–3.

Williams, A. 1962: The Barr and Lower Ardmillan Series (Caradoc) of the Girvan District, Scotland. *Geological Society, London, Memoirs 3*, 1–267, pls 1–25.

Williams, A. 1974: Ordovician Brachiopoda from the Shelve District, Shropshire. *Bulletin of the British Museum (Natural History), Geology Supplement 11*, 1–163, pls 1–28.

Williams, A. & Harper, D. A. T. 2000: Orthida. *In* Kaesler, R. L. (ed.): *Treatise on Invertebrate Paleontology, Vol. H (Brachiopoda) revised*, 714–782. Geological Society of America, Boulder and the Kansas University Press, Lawrence.

Wright, A. D. & Jaanusson, V. 1993: New genera of Upper Ordovician triplesiid brachiopods from Sweden. *Geologiska Föreningens i Stockholm Förhandlingar 115*, 93–108.

Xia, F., Zhang, S. & Wang, Z. 2007: The Oldest Bryozoans: New Evidence From The late Tremadocian (Early Ordovician) of east Yangtze gorges in China. *Journal of Paleontology 81*, 1308–1326.

Xu, H. 1979: Brachiopoda. *In* Jin, Y., Ye, S., Xu, H. & Sun, D. (eds): *Palaeontological atlas of north-western China*. 1, Qinghai, 60–112. Geological Publishing House, Beijing, 393 pp. [In Chinese].

Zeng, Q. 1987: Brachiopods. *In* Wang, X., Ni, S., Zheng, Q., Xu, G., Zhou, T., Li, Z., Xian, L. & Lai, C. (eds.): *Biostratigraphy of the Yangtze Gorge area: Early Palaeozoic Era*, 209–244. Geological Publishing House, Beijing, 614 pp. [In Chinese].

Zhan, R. & Cocks, L. R. M. 1998: Late Ordovician brachiopods from South China and their palaeogeographical significance. *Special Papers in Palaeontology 59*, 1–70, pls 1–9.

Zhan, R. & Jin, J. 2014: Early–Middle Ordovician brachiopod dispersal patterns in South China. *Integrative Zoology 9*, 121–140.

Zhan, R., Jin, J., Rong, J. & Liang, Y. 2013: The earliest known strophomenoids (Brachiopoda) from early Middle Ordovician rocks of South China. *Palaeontology 56*, 1121–1148.

Zhan, R., Li, R. Y., Percival, I. G. & Liang, Y. 2011: Brachiopod biogeographic change during the Early to Middle Ordovician in South China. *Memoirs of the Association of Australasian Palaeontologists 41*, 273–287.

Zhan, R. & Rong, J. 1995: Four new Late Ordovician brachiopod genera from the Zhejiang-Jiangxi border region, east China. *Acta Palaeontologica Sinica 34*, 549–574, pls 1–4.

Zhu, C. 1985: Late Ordovician Brachiopoda from Wuluntun district of eastern Da Hinngan Region, north-east China. *Bulletin of the Shenyang Institute of Geology and Mineral Resources, Chinese Academy of Sciences 12*, 24–54, pls 1–10. [In Chinese].

PLATES 1 – 21

Plate 1

1:	*Meristopacha* sp.
	Ventral exterior. NMW 98.28G.2162.
	Locality 8122, Berkutsyur Formation (Sandbian).
	Scale bars 2 mm.
2–11, 15, 16:	*Aploobolus tenuis* n. gen., n. sp.
	All from Locality 1022, Berkutsyur Formation (Sandbian).
2, 5:	Holotype, ventral internal mould, enlarged view of ventral pseudointerarea. NMW 98.28G.2135.
3, 6:	Dorsal internal mould, enlarged view of dorsal pseudointerarea. NMW 98.28G.2136.
4:	Latex cast of dorsal exterior. NMW 98.28G.2138.
7:	Disarticulated shell, ventral external mould and dorsal internal mould. NMW 98.28G.2134.1-2.
8:	Latex cast of disarticulated shell, ventral exterior and dorsal interior moulds; NMW 98.28G.2134.1-2.
9:	Dorsal exterior. NMW 98.28G.2144.
10:	Dorsal exterior. NMW 98.28G.2139
11:	Latex cast of ventral exterior. NMW 98.28G2143.
15:	Latex cast of dorsal exterior. NMW 98.28G.2140.
16:	Dorsal exterior. NMW 98.28G.2138.
2–5:	Scale bars 2 mm.
6–11, 15, 16:	Scale bars 1 mm.
12:	*Oxlosia* sp.
	Tastau Formation (upper Darriwilian), Locality 154.
	Dorsal exterior. NMW 98.28G.2128.
	Scale bar 2 mm.
13, 14:	*Broeggeria* cf. *B. putilla* (Tenjakova, 1989)
	Tastau Formation (late Darriwilian), Locality 163.
13:	Slightly exfoliated ventral exterior showing pseudointerarea. Locality 162. NMW 98.28G.2129.
	Scale bars 2 mm.
14:	Dorsal exterior. NMW 98.28G.2130.

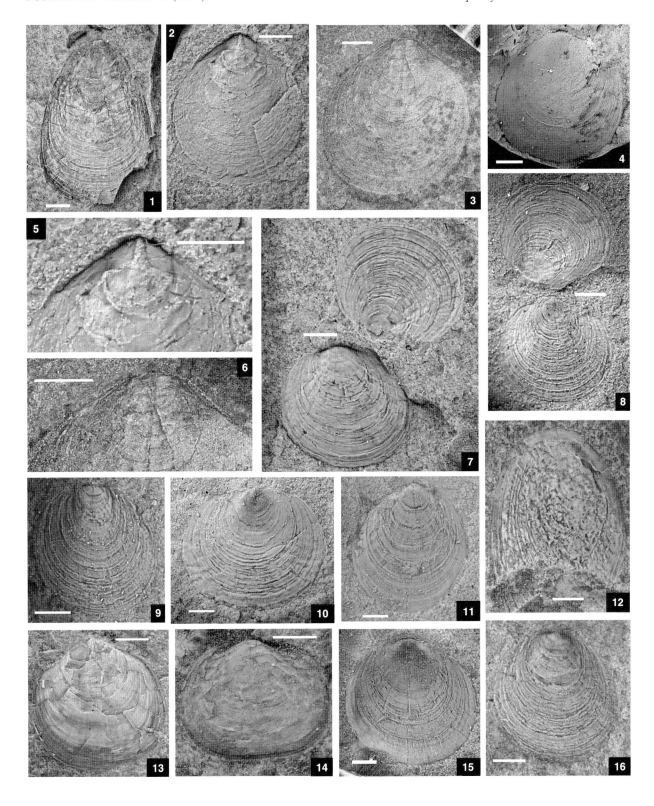

Plate 2

1:   *Petrocrania* **sp**.
                 Kopkurgan Formation (Katian), Locality 735
                 Dorsal internal mould. BC 63822b.
                 Scale bar 2 mm.

2–4:   *Eichwaldia* **sp**.
                 Berkutsyur Formation (Sandbian), Locality 8121, NMW 98.28G.2110,
                 Ventral valve, dorsal view, enlarged view of surface ornament, oblique posterior view;
                 2, 4, scale bars 500 μm; 3, scale bar 100 μm.

5, 7:   *Trematis*? **sp**.
                 Berkutsyur Formation (Sandbian), Locality 8234; BC 63865a.
                 Dorsal external mould, enlarged surface ornament.
                 5, scale bar 2 mm; 7, scale bar 1 mm.

6:   *Trematis* **aff**. *T. parva* **Cooper**, **1956**
                 Baigara Formation (late Darriwilian – early Sandbian), Locality 1026b, CNIGR 214/11352. Ventral exterior.
                 Scale bar 2 mm.

8:   *Ectenoglossa sorbulakensis* **Popov**, **1980**
                 Baigara Formation (late Darriwilian – early Sandbian), Locality 1021; NMW 98.28G.2133. Ventral exterior.
                 Scale bar 2 mm.

9–10:   *Elliptoglossa magna* **Popov**, **1977**
                 Kopkurgan Formation (late Sandbian).
9:   Ventral exterior. Locality 8122; BC 62411.
10:   Dorsal exterior. Locality 8236; NMW 98.28G.2132,
                 Scale bars 1 mm.

11–16:   *Meristopacha* **sp**.
                 Kopkurgan Formation (Sandbian), Locality 388.
11:   Dorsal exterior. NMW 98.28G.2145.
12:   Ventral exterior. NMW 98.28G.2146.
13:   Dorsal exterior. NMW 98.28G.2147.
14:   Dorsal internal mould. NMW 98.28G.2151.
15:   Dorsal exterior. NMW 98.28G.2148.
16:   Ventral internal mould. NMW 98.28G.2150.
                 Scale bars 1 mm.

17:   *Esilia* **cf**. *E. tchetverikovae* **Nikitin & Popov**, **1985**
                 Kipchak Limestone (late Darriwilian – early Sandbian), Locality 166.
                 Ventral internal mould. NMW 98.28G.2111.
                 Scale bar 1 mm.

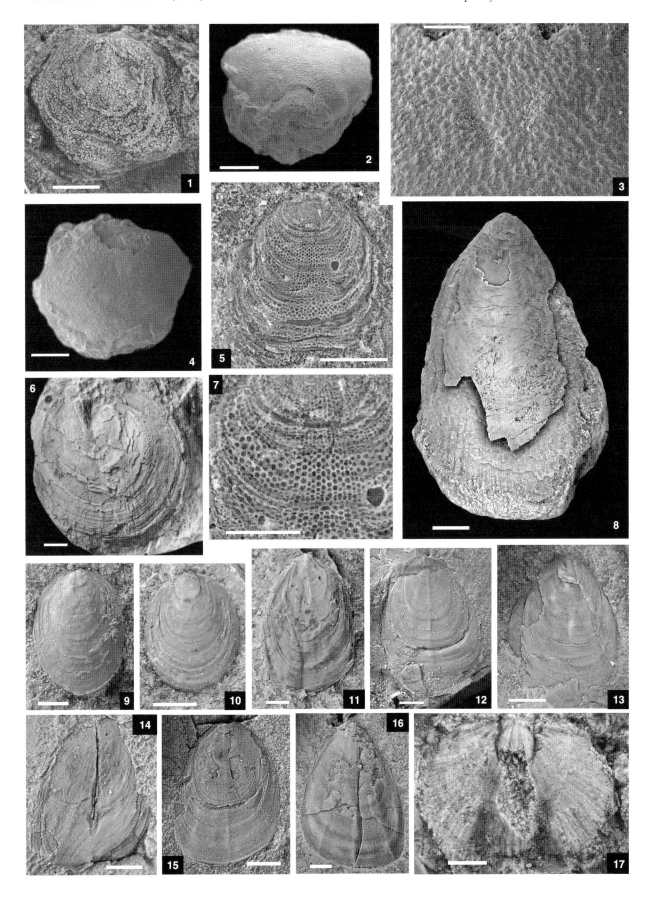

Plate 3

1–5:	***Esilia* cf. *E. tchetverikovae* Nikitin & Popov, 1985**
	Berkutsyur Formation (Sandbian), Locality 8233.
1, 2:	Dorsal exterior, dorsal and side views. BC 62425.
3–5:	Conjoined valves ventral, dorsal and anterior views. BC 62426.
	Scale bars 2 mm.
6–8:	**Furcitellinae gen. et sp. indet.**
	Baigara Formation (late Darriwilian – early Sandbian), Locality 1022.
6:	Conjoined valves, dorsal view. NMW 98.28G.1039.
7:	Conjoined valves, ventral view. NMW 98.28G.1040.
8:	Conjoined valves, dorsal view. NMW 98.28G.1041.
	Scale bars 2 mm.
9–16:	***Testaprica alperovichi* n. sp.**
	Berkutsyur Formation (Sandbian), Locality 813.
9:	Ventral exterior. NMW 98.28G.1228.
10:	Latex cast of dorsal exterior. BC 62342.
11, 16:	Latex cast of interior and dorsal internal mould. BC 62343.
12, 13:	Latex cast of dorsal exterior, dorsal and side views. BC 62341.
14, 15:	Holotype, ventral internal mould and latex cast of it. BC 62344.
	Scale bars 2 mm.
17–23:	***Colaptomena insolita* (Nikitina, 1985)**
	17, 19: Rgaity Formation (late Darriwilian), North Tien Shan, southern Kendyktas Range, Tolapty River; 18, 20–23: Baigara Formation (late Darriwilian – early Sandbian).
17:	Ventral exterior. Locality 11221; NMW 98.66G.2113.
18:	Conjoined valves, dorsal view. Locality 1021; NMW 98.28G.2116,
19:	Ventral internal mould. Locality 11221; NMW 98.66G.2112.
20, 21:	Conjoined valves, ventral and dorsal views. Locality 1021; NMW 98.28G.1378.
22, 23:	Conjoined valves, ventral and dorsal views. Locality 1022; NMW 98.28G.924.
	Scale bars 2 mm.

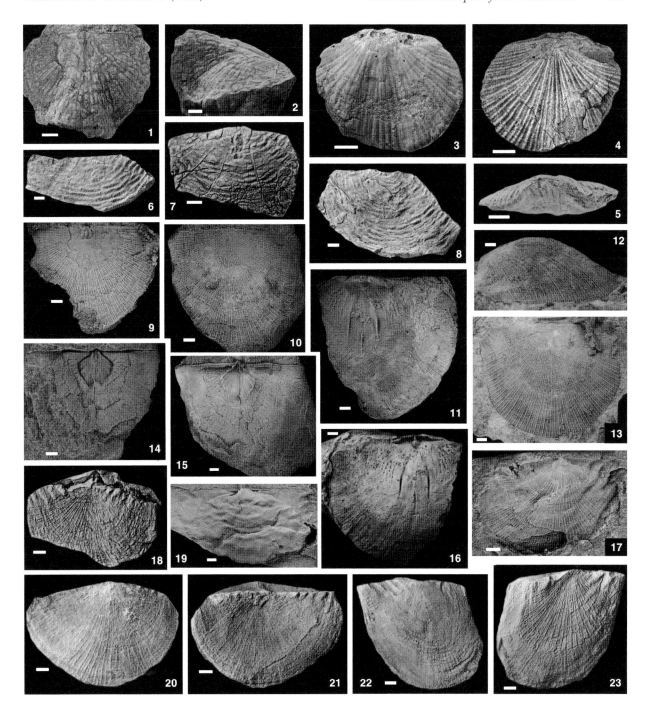

Plate 4

1:	***Rhipidomena* sp**.

              Kopkurgan Formation (Katian).
              Latex cast of ventral internal mould. Locality 735; BC 63820f.
              Scale bar 2 mm.

2–9:　　　　　***Katastrophomena rukavishnikovae* (Nikitina, 1985)**
              Baigara Formation (late Darriwilian – early Sandbian).
2:　　　　　　Ventral view of conjoined valves. Locality 1022; NMW 98.28G.861.
3, 4:　　　　Conjoined valves ventral and dorsal views. Locality 1022; NMW 98.28G.860.
5:　　　　　　Ventral interior. Locality 1021, NMW 98.28G.863.
6, 8:　　　　Conjoined valves ventral view, enlarged radial ornament. Locality 1021, NMW 98.28G.864,
7, 9:　　　　Conjoined valves ventral and dorsal views. Locality 1022; NMW 98.28G.862.
              Scale bars 2 mm

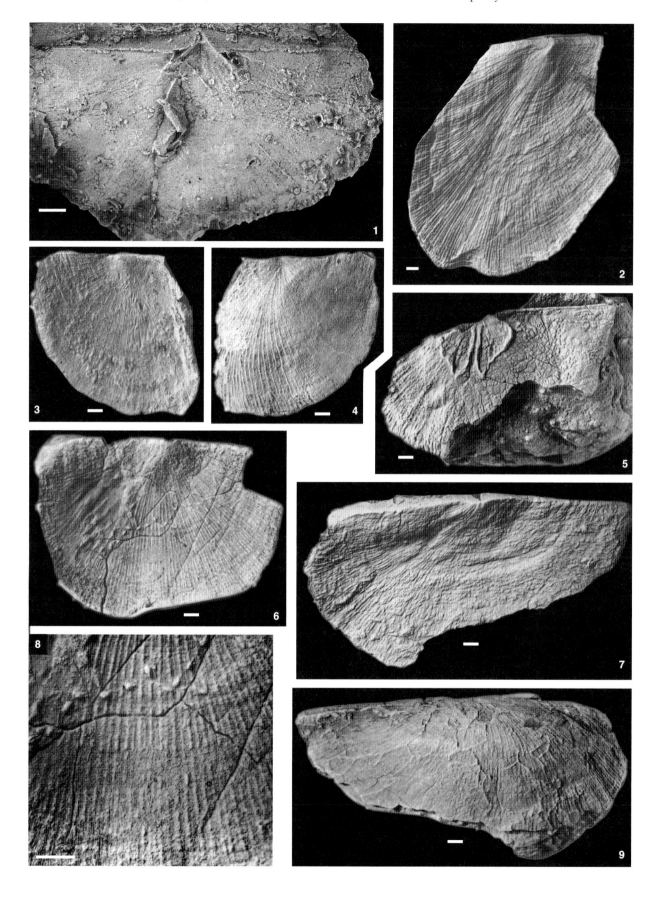

Plate 5

1–11:	***Doughlatomena splendens* n. gen., n. sp**.
	Kopkurgan Formation (Katian), Locality 735.
1:	Ventral exterior. BC 63888d.
2:	Ventral internal mould. BC 63849.
3, 6:	Ventral and lateral views of ventral internal mould. BC 63823.
4, 5:	Holotype, latex cast and dorsal internal mould. BC 63873a.
7, 8:	Latex cast and dorsal internal mould. BC 63847c.
9:	Ventral exterior. BC 63832.
10:	Dorsal internal mould. BC 63824b.
11:	Latex cast of ventral external mould. BC 63824c.
	Scale bars 5 mm.

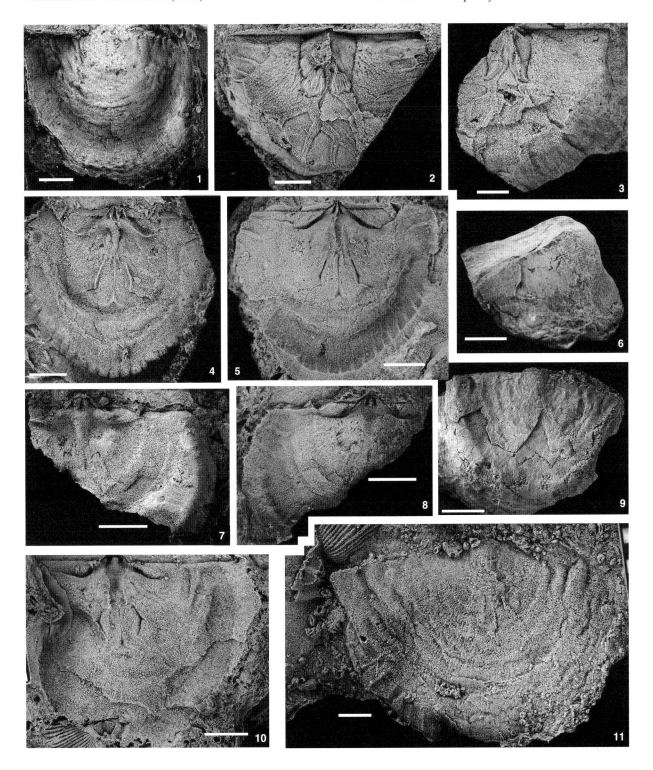

Plate 6

1, 2:	***Bandaleta* sp**.
	Berkutsyur Formation (Sandbian), SW of Lake Alakol.
1:	Dorsal exterior. Locality 8124; BC 62422.
2:	Ventral exterior. Locality 816, NMW 98.28G.2057.
	Scale bars 2 mm.

3–8:	***Isophragma princeps* Popov, 1980a**
	Berkutsyur Formation (Sandbian).
3:	Ventral exterior. Locality 812; BC 62332.
4:	Latex cast of ventral interior. Locality 813; BC 62349a.
5:	Dorsal exterior. Locality 812; BC 62333.
6, 7:	Latex cast of dorsal interior and internal mould. Locality 813; BC 62350.
8:	Ventral internal mould. Locality 813; BC 60824.
	Scale bars 2 mm.

9–19:	***Bimuria karatalensis* n. sp**.
	9–11, 13–15, 17–19, Baigara Formation (late Darriwilian – early Sandbian); 12, 16, Berkutsyur Formation (Sandbian).
9:	Latex cast of dorsal exterior. Locality 1026; NMW 98.28G.1047.
10:	Holotype, ventral internal mould. Locality 1026; NMW 98.28G.1122.
11:	Latex cast of dorsal interior. Locality 1026; NMW 98.28G.1046.
12, 16:	Ventral and dorsal views of conjoined valves. Locality 816; BC 62379.
13, 14:	Conjoined valves, ventral and dorsal views. Locality 1026b; NMW 98.28G.1125.
15:	Dorsal interior. Locality 1026b, NMW 98.28G.1121.
17, 18:	Dorsal and ventral views of conjoined valves. Locality 1026b; NMW 98.28G.1126.
19:	Ventral internal mould. Locality 1026; NMW 98.28G.1123.
	Scale bars 2 mm.

20:	***Cooperea* sp**.
	Berkutsyur Formation (early Sandbian).
	Ventral exterior. Locality 8121; BC 62400.
	Scale bars 2 mm.

21–23:	***Bimuria* sp**.
	Baigara Formation (late Darriwilian – early Sandbian).
21:	Ventral exterior. Locality 1021; NMW 98.28G.2034.
22:	Dorsal view of conjoined valves. Locality 625/3; NMW 98.28G.2035.
23:	Polished section of articulated valves parallel to commissural plane. Locality 1023; NMW 98.28G.1985.
	Scale bars 2 mm.

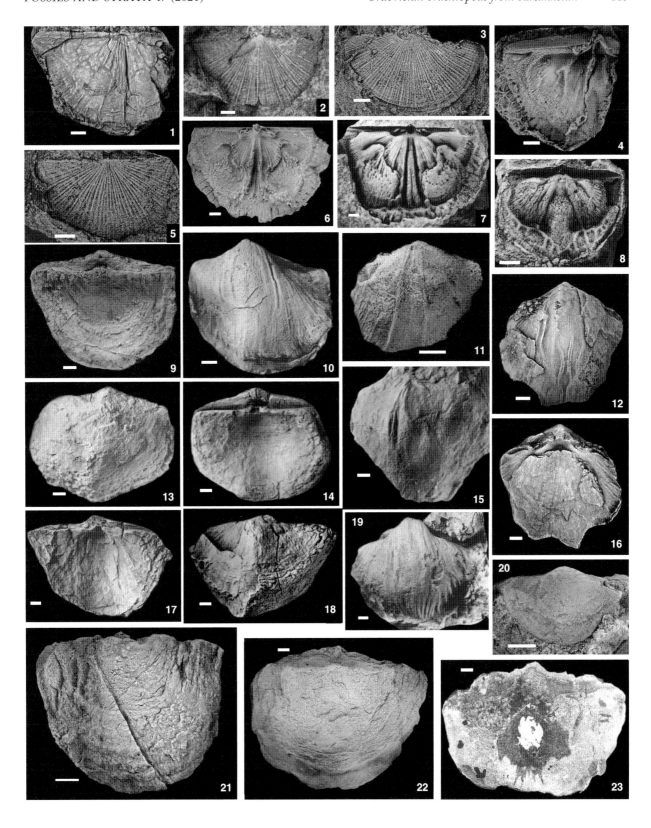

Plate 7

1–12, 15:   ***Acculina acculica* Misius *in* Misius & Ushatinskaya, 1977**
1–4: Baigara Formation (late Darriwilian – early Sandbian), Locality 1021;
5–8: Rgaity Formation (late Darriwilian), Tolapty River, southern Kendyktas Range, North Tien Shan, Locality 11221; 6–12, 15: Berkutsyur Formation (early Sandbian).

1: Ventral interior. NMW 98.28G.1129.
2, 3: Conjoined valves ventral and dorsal views. BC 62378.
4: Dorsal interior. NMW 98.28G.1130.
5, 8: Latex cast of interior and dorsal internal mould. NMW 98.66G.2117.
6, 7: Ventral internal mould and latex cast of it. NMW 98.66G.2118.
9: Latex cast of ventral exterior. Locality 812, BC 62322.
10: Latex cast of dorsal exterior. Locality 812, BC 62324.
11: Dorsal valve. Locality 812; BC 60874.
12: Latex cast of dorsal exterior. Locality 812; BC 62323.
15: Latex cast of dorsal interior. Locality 815, BC 62369.
Scale bars 2 mm.

13:   ***Dulankarella larga* Popov, Cocks & Nikitin, 2002**
Berkutsyur Formation (early Sandbian).
Latex cast of dorsal exterior. Locality 815; BC 62363.
Scale bar 2 mm.

14:   ***Bandaleta* sp.**
Berkutsyur Formation (early Sandbian).
Dorsal internal mould showing double septum and cardinalia. Locality 816; NMW 98.28G.2165.
Scale bar 2 mm.

16–18, 20–22:   ***Leptellina* sp.**
Baigara Formation (late Darriwilian – early Sandbian).
16, 17: Conjoined valves, ventral and dorsal views. Locality 1023; NMW 98.28G.1944.
18: Dorsal interior. Locality 1021; NMW 98.28G.1894.
20, 21: Conjoined valves, ventral and dorsal views. Locality 1023; NMW 98.28G.1892.
22: Ventral internal mould. Locality 1021; NMW 98.28G.1893.
Scale bars 2 mm.

19, 23:   ***Ptychoglyptus* sp.**
19: Kipchak Limestone (late Darriwilian–early Sandbian): 23: Berkutsyur Formation (early Sandbian).
19: Ventral valve. Locality 166; BC 62443.
23: Ventral valve. Locality 8121; BC 62403.
Scale bars 2 mm.

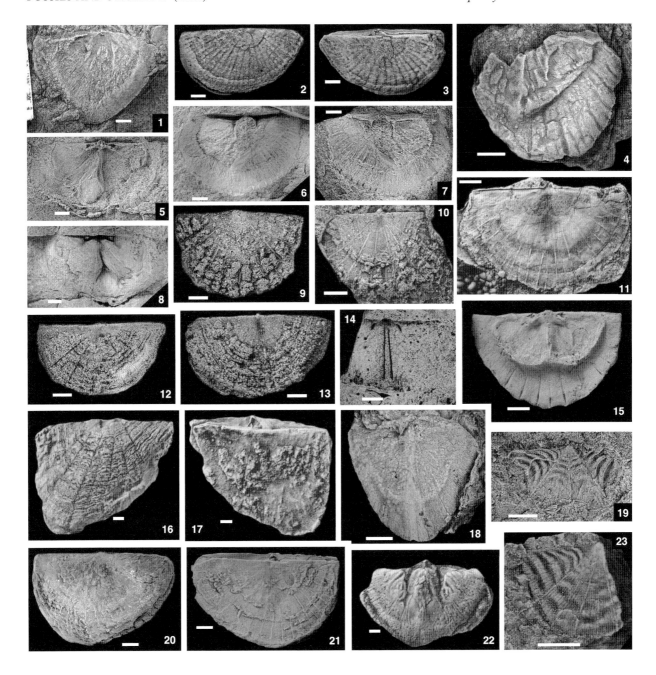

Plate 8

1–5:	***Dulankarella larga* Popov, Cocks & Nikitin, 2002**
	Berkutsyur Formation (early Sandbian).
1–3:	Conjoined valves lateral, dorsal and ventral views. Locality 817; BC 62390.
4:	Ventral internal mould. Locality 816; BC 62377.
5:	Ventral internal mould. Locality 816; BC 62376.
	Scale bars 2 mm.
6:	***Shlyginia* sp.**
	Kopkurgan Formation (early Sandbian).
	Ventral internal mould. Locality 8235; NMW 98.28G.2163.
	Scale bar 2 mm.
7–17:	***Kajnaria derupta* Nikitin & Popov *in* Klenina *et al.*, 1984**
	Baigara Formation (late Darriwilian – early Sandbian).
7, 8:	Ventral and posterior views of articulated shell. Locality 1026b; NMW 98.28G.2000.
9, 10:	Latex cast of ventral interior and internal mould. Locality 1026; NMW 98.28G.1169.
11:	Dorsal external mould and ventral interarea. Locality 1026; BC 12931.
12, 13:	Conjoined valves, ventral and posterior views. Locality 1026b; NMW 98.28G.2002.
14-17:	Conjoined valves, lateral, dorsal, ventral and posterior views. Locality 1026b; NMW 98.28G.2001.
	Scale bars 2 mm.
18–22:	***Apatomorpha akbakaiensis* n. sp.**
	Baigara Formation (late Darriwilian – early Sandbian);
18–20:	Conjoined valves, ventral, dorsal and lateral views. Locality 1022; NMW 98.28G.1239.
21:	Ventral internal mould. Locality 1023; NMW 98.28G.1241.
22:	Ventral internal mould. Locality 1022; NMW 98.28G.1242.
	Scale bars 2 mm.

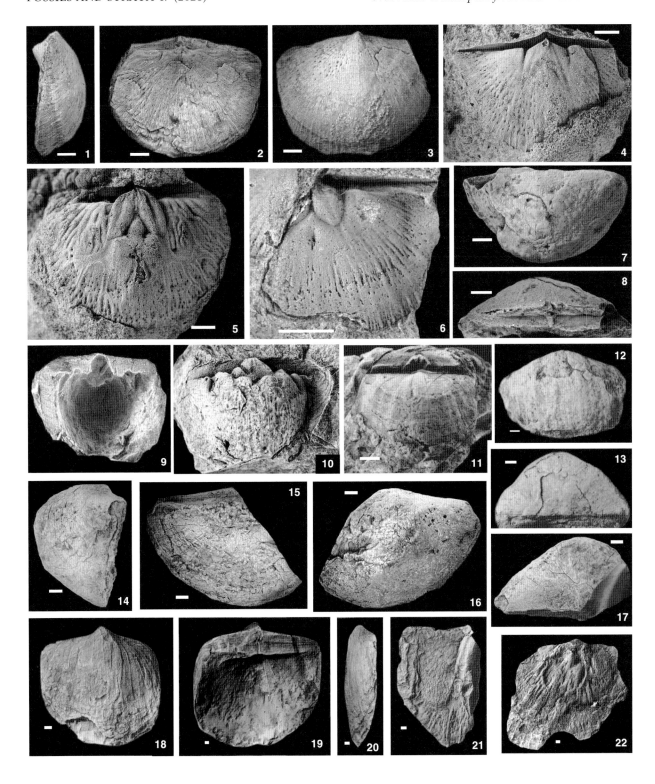

Plate 9

1–4:	***Apatomorpha akbakaiensis* n. sp**.

Baigara Formation (late Darriwilian – early Sandbian);

1-3: Holotype, articulated shell, dorsal, ventral and lateral views. Locality 1022; NMW 98.28G.1240.

4: Articulated shell, polished section parallel to the commissural plane (cp, cardinal process; s, median septum, sr, socket ridge and t, teeth). Locality 1023; NMW 98.28G.1243.
Scale bars 2 mm.

5–10, 13: ***Ishimia* aff. *I. ishimensis* Nikitin, 1974**
Berkutsyur Formation (early Sandbian).
5–7, Locality 8124, 8–10, 13, Locality 813.

5: Latex cast of dorsal exterior. BC 60732.
6: Latex cast of ventral exterior. BC 62417.
7: Latex cast of dorsal exterior. BC 62416.
8: Latex cast of dorsal exterior. BC 62418.
9: Exfoliated dorsal interior showing cardinal process and socket plates. NMW 98.28G.1984.
10: Ventral internal mould. BC 62353.
13: Ventral internal mould. BC 62352.
Scale bars 2 mm.

11, 12, 14, 15: ***Titanambonites* cf. *T. magnus* Nikitin, 1974**
Baigara Formation (late Darriwilian – early Sandbian).

11, 14: Ventral and dorsal views of internal mould. Locality 1026b, NMW 98.28G.1987.
12: Ventral internal mould of juvenile specimen, Locality 1026; BC 62447.
15: Ventral internal mould, Locality 1026b, NMW 96.28G.2109.
Scale bars 2 mm.

16: ***Anoptambonites kovalevskii* Popov, Nikitin & Cocks, 2000**
Kopkurgan Formation (Katian), latex cast of ventral interior, Locality 735, BC 63837a.
Scale bars 2 mm.

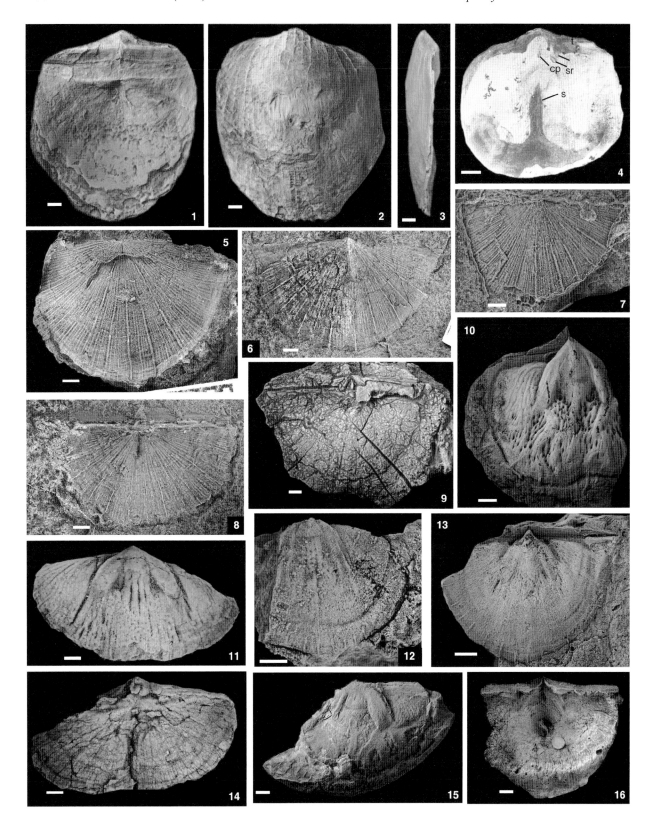

Plate 10

1–9:   ***Lepidomena betpakdalensis* n. sp**.
                 Baigara Formation (late Darriwilian – early Sandbian).
1:               Ventral interior. Locality 1023; NMW 98.28G.1289.
2, 3:   Articulated shell, dorsal and ventral views. Locality 1022; NMW 98.28G.1323.
4–6:   Holotype, articulated shell, ventral, lateral and dorsal views. Locality 1022; NMW 98.28G.1324.
7, 8:   Articulated shell, dorsal and ventral views; Locality 1022; NMW 98.28G.1325
9:               Dorsal interior. Locality 1022; NMW 98.28G.1326.
                 Scale bars 2 mm.

10, 11:   ***Tenuimena* aff**. ***T. planissima* Nikitina *et al*. 2006**
                 Kopkurgan Formation (early Sandbian), Locality 8122.
10:              Latex cast of disarticulated ventral exterior and dorsal interior. BC 62405a, b.
11:              Latex cast of dorsal exterior. BC 62406.
                 Scale bars 2 mm.

12:   ***Tesikella necopina* (Popov, 1980b)**
                 Berkutsyur Formation, Locality 8235.
                 Ventral exterior. BC 62436.
                 Scale bar 2 mm.

13–16:   ***Shlyginia extraordinaria* (Rukavishnikova, 1956)**
                 Unnamed formation (early to mid Katian), near Ergibulak well, North Betpak-Dala, Locality 1501.
13, 14:   Dorsal internal mould, latex cast of interior. NMW 98.28G.1961.
15:              Latex cast of ventral exterior. NMW 98.28G.1960.
16:              Ventral internal mould. NMW 98.28G.1962.
                 Scale bars 2 mm.

17, 18:   ***Sowerbyella (S.) ampla* (Nikitin & Popov, 1996)**
                 Unnamed formation (early to mid Katian), near Ergibulak well, North Betpak-Dala, Locality 1501.
17:              Dorsal internal mould. NMW 98.28G.2065.
18:              Dorsal internal mould. NMW 98.28G.2066.
                 Scale bars 2 mm.

19:   ***Qilianotryma* cf. *Q. suspectum* (Popov *in* Nikiforova *et al*., 1982)**
                 Unnamed formation (early to mid Katian), near Ergibulak well, North Betpak-Dala, Locality 1501.
                 Dorsal internal mould. NMW 98.28G.1966.
                 Scale bar 2 mm.

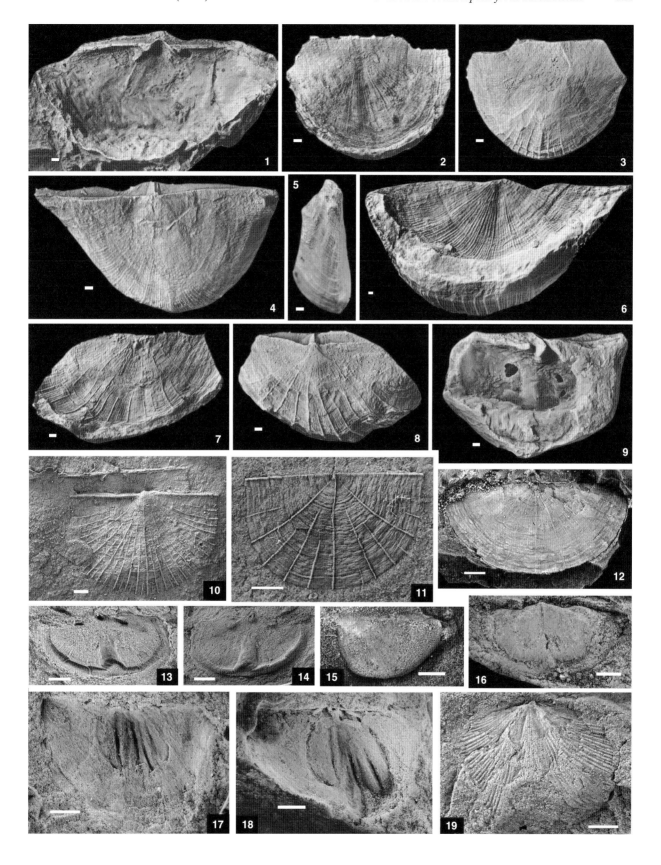

Plate 11

1–6:   *Kassinella simorini* **n. sp**.
         Berkutsyur Formation (early Sandbian), 7 km SW of Lake Alakol, Locality 8235.
1–4:   Holotype, disarticulated shell, latex cast of dorsal and ventral exterior, internal moulds of dorsal and ventral valves, side view of dorsal and ventral interior (latex cast), enlarged latex cast of dorsal and ventral interior. NMW 98.28G.1983.
5, 6:   Ventral internal mould and latex cast of it. NMW 98.28G.1980.
         Scale bars 2 mm.

7–10:   *Chonetoidea*? **sp**.
         Kopkurgan Formation (early Sandbian).
7:          Latex cast of dorsal exterior. Locality 8122; BC 62737.
8:          Latex cast of ventral valve exterior. Locality 8122; BC 62407.
9:          Latex cast of dorsal valve plus ventral interarea. Locality 8236a; BC 62441.
10:        Latex cast of dorsal exterior. Locality 8236; BC 62439.
         Scale bars 2 mm.

11, 13, 14:   *Akadyria simplex* **Nikitina, Neuman, Popov & Bassett, 2006**
         Tastau Formation (upper Darriwilian), Locality 154.
11:        Dorsal and ventral internal moulds of disarticulated shell. NMW 98.28G.1145.
13, 14:   Dorsal external mould and latex cast of dorsal exterior. NMW 98.28G.1146.
         Scale bars 2 mm.

12:   *Anoptambonites kovalevskii* **Popov, Nikitin & Cocks, 2000**
         Kopkurgan Formation (Katian), Locality 735.
         Dorsal internal mould of juvenile individual. BC 63886.
         Scale bar 2 mm.

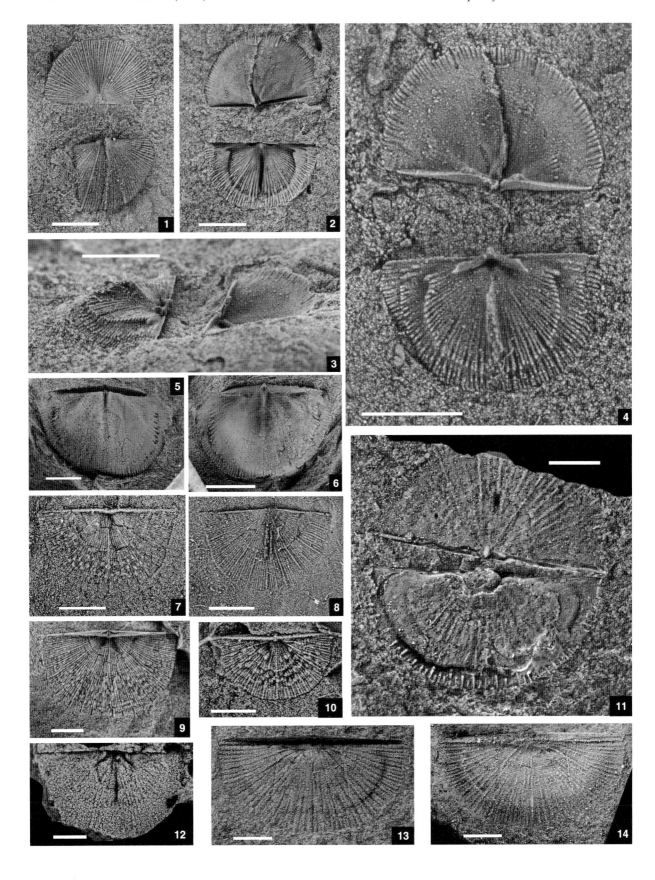

Plate 12

1:	***Gacella* sp**.
	Kopkurgan Formation (Katian), Locality 735.
	Ventral internal mould. BC 63822c.
	Scale bar 2 mm.
2–6, 11:	***Gunningblandella* sp**. **1**.
	Kopkurgan Formation (Katian), Locality 735.
2:	Ventral internal mould. BC 63830c.
3:	Ventral internal mould. BC 63829.
4:	Ventral internal mould; BC 63840a.
5:	Latex cast of dorsal external mould. BC 63837c.
6:	Latex cast of ventral external mould. BC 63874a.
11:	Ventral external mould. BC 63837d.
	Scale bars 2 mm.
7–10, 12–16:	***Sowerbyella* (*S.*) *ampla* (Nikitin & Popov, 1996)**
	Kopkurgan Formation (Katian), Locality 735.
7:	Ventral internal mould. BC 63840b.
8, 9:	Latex cast of dorsal interior and dorsal internal mould. BC 63838e.
10:	Ventral internal mould with superimposed partly preserved dorsal internal mould. BC 63882a.
12:	Latex cast of ventral exterior. BC 63824a.
13, 14:	Latex cast of dorsal interior and dorsal internal mould. BC 63844a.
15:	Dorsal view of conjoined valves, latex cast. BC 63844b.
16:	Latex cast of ventral valve exterior. BC 63839a.
	Scale bars 2 mm.

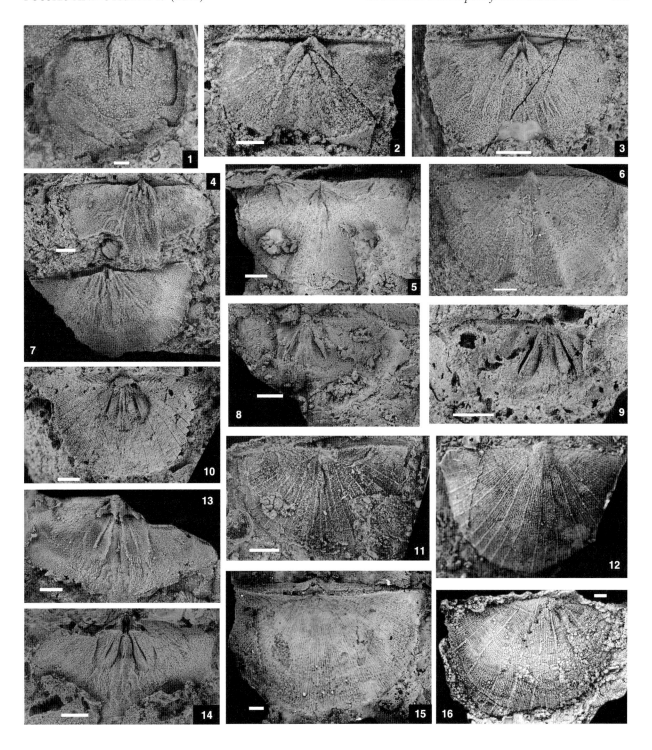

Plate 13

1:              ***Synambonites* sp**.
                Berkutsyur Formation (early Sandbian), Locality 8121.
                Dorsal exterior. BC 62404.
                Scale bar 2 mm.

2–17:   ***Sowerbyella (S.) verecunda baigarensis* n. subsp.**
                2, 3, 8, 9, 10, 14, 15, 17: Baigara Formation (late Darriwilian – early Sandbian)
                4, 7, 11–13, 16: Berkutsyur Formation (early Sandbian).
                5, 6: Takyrsu Formation, Kipchak Limestone (late Darriwilian – early Sandbian).
2:              Ventral internal mould. Locality 1028. NMW 98.28G.1150.
3:              Ventral internal mould. Locality 1026. BC 60862.
4:              Ventral internal mould. Locality 812, BC 62340.
5:              Ventral exterior. Locality 166, NMW 98.28G.1152.
6:              Dorsal exterior. Locality 166. NMW 98.28G.1153.
7:              Latex cast of dorsal interior. Locality 812. BC 62337.
8:              Latex cast of ventral exterior. Locality 1026. NMW 98.28G.1215.
9:              Ventral views of articulated shell. Locality 1021. NMW 98.28G.1222.
10:             Dorsal view of articulated shell. Locality 1021. NMW 98.28G.1223.
11:             Latex cast of dorsal exterior. Locality 815. NMW 98.28G.1999.
12:             Latex cast of ventral exterior. Locality 815. NMW 98.28G.1220.
13:             Latex cast of dorsal interior. Locality 812. BC 62326.
14:             Latex cast of dorsal exterior. Locality 812. BC 62325.
15:             Dorsal view of articulated shell. Locality 1021. NMW 98.28G.1224.
16:             Ventral internal mould. Locality 812. NMW 98.28G.1217.
17:             Dorsal internal mould. NMW 98.28G.1151.
                Scale bars 2 mm.

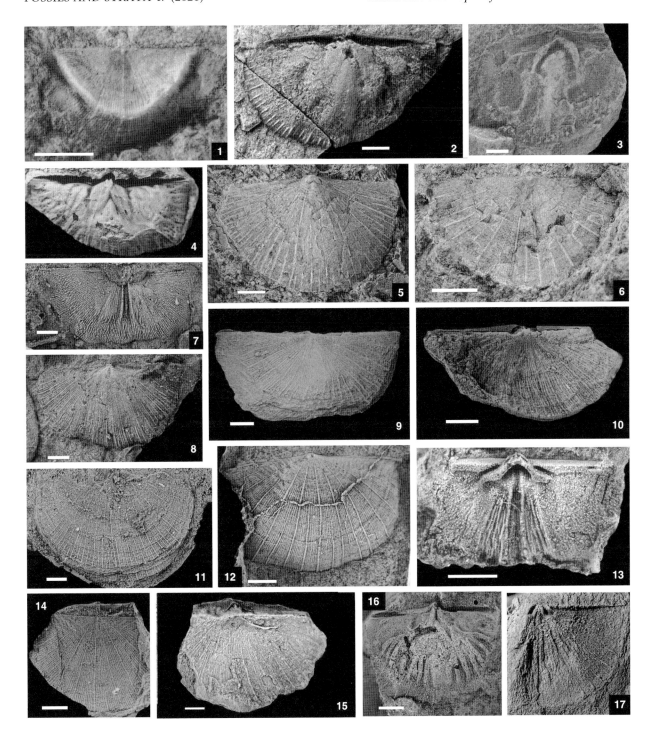

Plate 14

1, 2, 4, 5:	***Grammoplecia* aff. *G. globosa* (Nikitin & Popov, 1985)**
	Baigara Formation (late Darriwilian – early Sandbian), Locality 1023.
	Conjoined valves, dorsal, ventral, lateral and anterior views. NMW 98.28G.1382.
	Scale bars 2 mm.
3, 6–17:	***Grammoplecia globosa* (Nikitin & Popov, 1985)**
	Baigara Formation (late Darriwilian – early Sandbian).
3:	Ventral internal mould. Locality 1026. NMW 98.28G.1397.
6, 8, 9, 11:	Conjoined valves, anterior, ventral, dorsal and side views. Locality 1026b. NMW 98.28G.1398.
7, 10, 12:	Conjoined valves, ventral, dorsal and anterior views. Locality 1026b. NMW 98.28G.1399.
14–17:	Conjoined valves, ventral, lateral, dorsal and anterior views. Locality 1026b. NMW 98.28G.1400.
	Scale bars 2 mm
18–20:	***Triplesia* sp.**
	Berkutsyur Formation (early Sandbian), Locality 8121.
	Lateral, dorsal and ventral views. BC 63698.
	Scale bars 1 mm

Plate 15

1–5, 17:	***Bokotorthis kasachstanica* (Rukavishnikova, 1956)**
1–5:	1–5: Kopkurgan Formation (late Sandbian – mid Katian), Locality 735.
	17: unnamed formation (early to mid Katian), near Ergibulak well, North Betpak-Dala, Locality 1501.
1:	Ventral internal mould. BC 63846a.
2:	Dorsal internal mould. BC 63828,
3:	Latex cast of dorsal exterior. BC 63876a.
4:	Latex cast of ventral external mould. BC 63847a.
5:	Dorsal internal mould. BC 63877.
17:	Ventral external mould. NMW 98.28G. 2064.
	Scale bars 2 mm.

6, 7:   ***Dolerorthis expressa* Popov, 1980b**
Berkutsyur Formation (early Sandbian), Locality 817.
Latex cast of dorsal interior and dorsal internal mould. BC 62387.
Scale bars 2 mm.

8, 11, 15:   **Hesperorthidae gen. et sp. indet.**
Kipchak Limestone (late Darriwilian-early Sandbian), Locality 166.
8:    Ventral exterior. NMW 98.28G.2052.
11:    Exfoliated dorsal exterior. NMW 98.28G.2054.
15:    Ventral internal mould. NMW 98.28G.2055.
Scale bars 2 mm.

9, 10:   ***Lictorthis* cf. *L. licta* (Popov & Cocks, 2006)**
Kopkurgan Formation (late Sandbian – mid Katian), Locality 735.
Ventral internal mould and latex cast of ventral interior. BC 63874b.
Scale bars 2 mm.

12, 13:   ***Sowerbyella (S.)* cf. *S. acculica* Misius, 1986**
Kopkurgan Formation (upper Sandbian), Locality 816.
Ventral exterior, anterior and ventral views. BC 62373.
Scale bars 2 mm.

14:   ***Sowerbyella (S.) verecunda baigarensis* n. subsp.**
Baigara Formation (late Darriwilian – early Sandbian), Locality 1026.
Holotype, dorsal internal mould. NMW 98.28G.1214.
Scale bar 2 mm.

16:   ***Gunningblandella* sp. 2**
Anderken Formation (late Sandbian), Burultas Valley, Locality 538; BC 32727, ventral internal mould.
Scale bar 2 mm.

18:   ***Costistriispira?* sp.**
Kopkurgan Formation (upper Sandbian), Locality 8122.
Latex cast of dorsal valve. BC 62410.
Scale bars 2 mm.

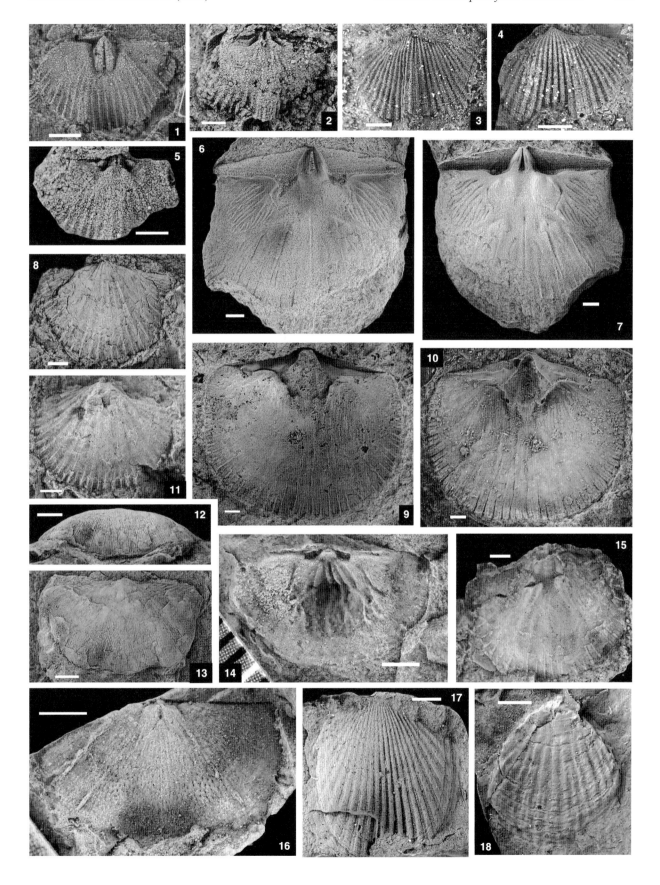

Plate 16

1:	***Scaphorthis recurva* Nikitina, 1985**
	Baigara Formation (late Darriwilian – early Sandbian), Locality1021.
	Ventral interior. NMW 98.28G.1872.
	Scale bars 2 mm.

2–7, 9:	***Sonculina* cf. *S. prima* Misius, 1986**
	Baigara Formation (late Darriwilian – early Sandbian), Locality 1026b.
2–4:	Articulated valves, dorsal, ventral and side views. NMW 98.28G.1147.
5, 9:	Internal mould of conjoined valves, dorsal and ventral views. NMW 98.28G.1148.
6, 7:	Articulated valves, dorsal and ventral views. NMW 98.28G.1149.
	Scale bars 2 mm.

8, 10–16, 18–21:	***Sonculina baigarensis* n. sp.**
	Baigara Formation (late Darriwilian – early Sandbian).
8:	Dorsal interior. Locality 1021. NMW 98.28G.1846.
10, 11:	Articulated valves, dorsal and ventral views. Locality 1022. NMW 98.28G.2121.
12–14:	Articulated valves, ventral, dorsal and side views. Locality 1022. NMW 98.28G.1888.
15, 16, 18:	Articulated valves, dorsal, ventral and side views. Locality 625/3. NMW 98.28G.1890.
19–21:	Holotype, articulated valves, ventral, dorsal and side views. Locality 625/3. NMW 98.28G.1889.
	Scale bars 2 mm.

17, 22, 27:	***Plectocamara* sp.**
	Baigara Formation (late Darriwilian – early Sandbian), Locality 1021.
	Anterior, dorsal and ventral view of articulated shell. NMW 98.28G.1981.
	Scale bars 2 mm.

23–26:	***Pionodema opima* Popov, Cocks & Nikitin, 2002**
	Kopkurgan Formation (early Sandbian), Locality 8235.
23:	Latex cast of dorsal exterior. NMW 98.28G.1970.
24:	Latex cast of dorsal interior. NMW 98.28G.1969.
25, 26:	Latex cast of ventral interior and ventral internal mould. NMW 98.28G.1967.
	Scale bars 1 mm.

Plate 17

1–12:	*Eridorthis* **sp**.
	1–8, 9–12: Berkutsyur Formation (early Sandbian), Locality 8233.
	9: Kipchak Limestone (late Darriwilian – early Sandbian), Locality 166.
1, 5:	Ventral and dorsal views of exfoliated shell showing ventral muscle field. NMW 98.28G.2010.
2, 8, 10:	Articulated valves ventral, dorsal and side views. NMW 98.28G.2011.
3, 7, 11:	Articulated valves ventral, dorsal and side views. NMW 98.28G.2012.
4, 8, 12:	Exfoliated complete shell ventral, dorsal and side views. NMW 98.28G.2013
9:	Ventral exterior. NMW 98.28G.1997.
	Scale bars 2 mm.

13–19:	*Altynorthis betpakdalensis* **n. sp**.
	Baigara Formation (late Darriwilian – early Sandbian).
13–15:	Conjoined valves, dorsal, ventral and side views. Locality 1021. NMW 98.28G.1468.
16, 19:	Conjoined valves, dorsal and ventral views. Locality 1023. NMW 98.28G.1464, 1465.
17:	Dorsal internal mould. Locality 1023. NMW 98.28G.1466.
18:	Ventral internal mould. Locality 1021. NMW 98.28G.1467.
	Scale bars 2 mm.

Plate 18

1–8:	***Altynorthis betpakdalensis*** **n. sp**.
	Baigara Formation (late Darriwilian – early Sandbian).
1–3:	Holotype, articulated shell, dorsal, side and ventral views. Locality 1021, NMW 98.28G.1469.
4:	Dorsal valve interior. Locality 1023. NMW 98.28G.1471.
5–7:	Articulated shell, dorsal, side and ventral views. Locality 1020a. NMW 98.28G.1470.
8:	Ventral valve interior. Locality 1022. NMW 98.28G.1472.
9–14:	***Altynorthis tabylgatensis*** **(Misius, 1986)**
	9–11: Baigara Formation (late Darriwilian – early Sandbian).
	12, 13: Berkutsyur Formation (early Sandbian), Locality 8234.
	14: Kipchak Limestone (late Darriwilian – early Sandbian), Locality 166.
9, 10:	Dorsal internal mould and latex cast of ventral interior. Locality 1026. BC 60833.
11:	Dorsal internal mould. Locality 1028. NMW 98.28G.1238.
12, 13:	Latex cast of ventral interior and ventral internal mould. BC 63866.
14:	Ventral exterior. NMW 98.28G.2060.
	Scale bars 2 mm

Plate 19

1–12:	***Altynorthis vinogradovae* n. sp**.
	Kopkurgan Formation (early Sandbian).
1:	Ventral internal mould. Locality 814. BC 60801.
2, 3:	Latex cast of ventral valve interior and ventral internal mould. Locality 813. BC 62345.
4:	ventral internal mould. Locality 813. BC 60826.
5, 6:	Holotype, latex cast of dorsal valve interior and dorsal internal mould. Locality 814. BC 62356
7:	Exfoliated dorsal exterior. Locality 813. BC 60828.
8:	Latex cast of ventral exterior. Locality 813. BC 62347.
9:	Dorsal internal mould. Locality 814. NMW 98.28G.1972.
10:	Latex cast of dorsal exterior. Locality 814. BC 62357.
11:	Latex cast of dorsal interior. Locality 813. BC 62346.
12:	Latex cast of dorsal exterior. Locality 813, BC 62348,
	Scale bars 2 mm

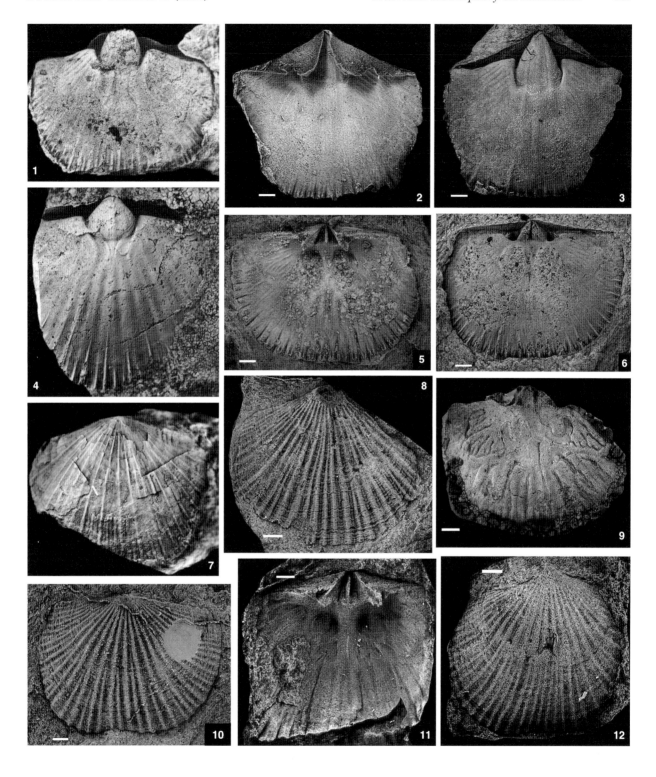

Plate 20

1, 2:	***Sonculina baigarensis* n. sp**.
	Baigara Formation (late Darriwilian – early Sandbian), Locality 1022.
	Dorsal and ventral views of internal mould. NMW 98.28G.2122.
3–10:	***Pionodema opima* Popov, Cocks & Nikitin, 2002**
	Kopkurgan Formation (early Sandbian).
3, 4:	Latex cast of dorsal interior and dorsal internal mould. Locality 8235. NMW 98.28G.1968.
5–8:	Ventral, dorsal, anterior and posterior views of conjoined valves. Locality 817. BC 62397.
9, 10:	Dorsal and ventral views of internal mould. Locality 817. BC 62398.
12–14, 16:	***Ilistrophina tesikensis* Popov, Cocks & Nikitin, 2002**
	Berkutsyur Formation (early Sandbian), Locality 8121.
	Ventral, dorsal, lateral, and anterior views of articulated shell. BC 62423.
11, 15, 17–19:	*Ancistrorhyncha modesta* **Nikiforova & Popov, 1981**
	Baigara Formation (late Darriwilian–early Sandbian), Locality N-12.
11, 15:	Dorsal and lateral views of dorsal exterior. NMW 98.28G.2009.
17:	Ventral exterior. NMW 98.28G.2007.
18:	Ventral exterior. NMW 98.28G.2008.
19:	Dorsal exterior. NMW 98.28G.2016.
	Scale bars 2 mm

Plate 21

1–6:	*Baitalorhyncha rectimarginata* **gen**. et **sp**. **nov**.
	Berkutsyur Formation (early Sandbian).
	1–4: Locality 8121; 5–16: Locality 8234.
1–4:	Holotype, ventral, dorsal, lateral, and anterior views of conjoined valves. BC 62402.
5:	Dorsal internal mould. BC 62386.
6:	Ventral exterior. BC 62392.
7:	Dorsal exterior. BC 62393.
8:	Ventral internal mould. BC 62391.
9:	Latex cast of ventral exterior. BC 63865b.
10:	Latex cast of dorsal exterior. BC 63853b.
11:	Latex cast of dorsal exterior. BC 63857b.
12:	Dorsal internal mould. BC 63857a.
13:	Ventral internal mould. BC 63854a.
14:	Ventral internal mould. BC 63854b.
15:	Ventral exterior. BC 63853a.
16:	Latex cast of dorsal exterior. BC 63859b.
	Scale bars 1 mm

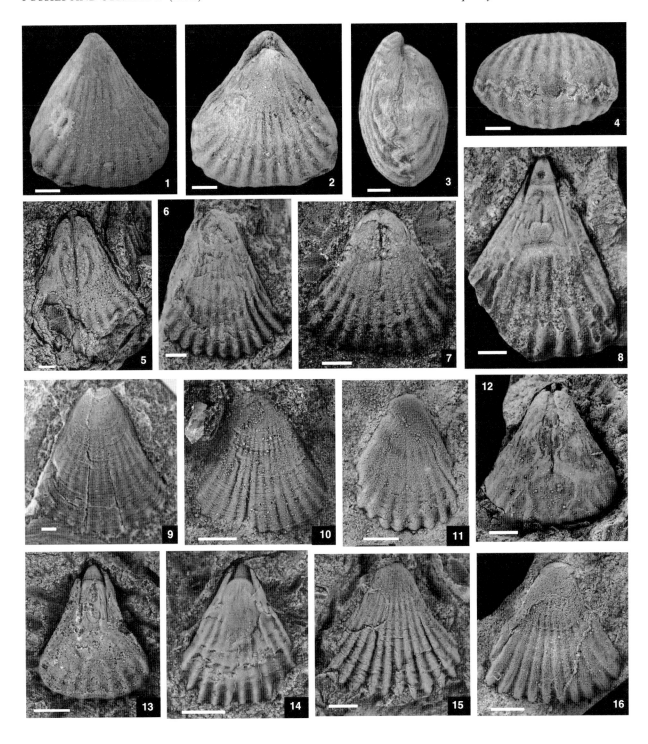

Appendix

The present paper finalises more than 60 years history of Ordovician brachiopod studies in South Kazakhstan which started with the pioneering publication by Rukavishnikova (1956). The comprehensive list of the Ordovician (Dapingian to Hirnantian) brachiopod taxa (Tables 1, 2) documented from the Chu-Ili Range, West Balkhash Region, and Betpak-Dala desert includes almost 300 brachiopod species, 167 of them identified to species level, while the others are in open nomenclature. It seems likely that the generic diversity of the Mid to Late Ordovician brachiopods from South Kazakhstan is now well established, whilst there is much still to be done at the species level, whose diversity appears to be outstanding. In particular, much more is needed to document the α and β diversity of the benthic faunas associated with various organic build-ups, which represent major biodiversity hot-spots in the region. However, the golden age of the brachiopod studies in Kazakhstan is largely over: the few active specialists left are of a good age, and a significant part of the brachiopod collections made by Russian and Kazakh geologists more than half a century ago has been lost or destroyed. However, we hope that the data on the fossil localities given in this paper can be used as a guide in assisting the new generations of Kazakh palaeontologists in gathering additional specimens and data for more advanced research.

The list brachiopod of taxa in Tables 1, 2 is based, as well as the present study, on publications by Rukavishnikova (1956), Sapelnikov & Rukavishnikova (1973, 1975), Nazarov & Popov (1980), Nikitin *et al.* (1980, 1996), Nikitin & Popov (1996), Nikitina (1985), Nikitina *et al.* (2006), Popov (1980b), Popov *et al.* (1999, 2000, 2002), and Popov & Cocks (2006). Generic and species discrimination of many taxa in those publications was reassessed and updated before inclusion here.

Table 1. Stratigraphical distribution of the Katian to Hirnantian brachiopod species at the West Balkhash Region, Chu-Ili Range and Betpak-Dala desert.

		Global stages				Lithostratigraphical units							
		Katian											
		Ka1	Ka2–Ka3	Ka4	Hirnantian	Anderken Fm.	Kopkurgan Fm.	Otar Mb	Degeres Mb.	Akkol Mb.	Sartan-Manai Lm.	Ulkuntas Fm.	Zhalair Fm.
1		2	3	4	5	6	7	8	9	10	11	12	13
Order LINGULIDA													
1	Acrosaccus aff. A. posteroconvexus	+									+		
2	Acrosaccus sp.		+						+				
3	Orbiculoidea sp.			+								+	
4	Paterula sp.			+								+	
5	Pseudolingula sp.	+									+		
6	Schizotreta triangularis	+									+		
7	Trematis sp. 4		+						+				
Order CRANIIDA													
8	Petrocrania sp.*		+				+						
Order CRANIOPSIDA													
9	Paracraniops sp. 2		+						+				
Order TRIMERELLIDA													
10	Adensa monstratum		+							+			
11	Eodinobolus kasachstanicus		+					+	+	+			
12	Eodinobolus sp.		+							+			
13	Palaeotrimerella medojevi		+							+			
Order STROPHOMENIDA													
14	Acculina? sp.*	+											
15	Aegiromena durbenensis		+		+		+						+
16	Anisopleuriella novemcostata				+								+
17	Anisopleurella sp.	+				+							
18	Anoptambonites kovalevskii*	+	+					+	+	+			
19	Anoptambonites subcarinatus	+									+		
20	Bandaleta plana	+	+								+		
21	Bandaleta sp.	+											
22	Bellimurina sp.	+	+					+		+	+		
23	Chonetoidea sp.	+				+							
24	Christiania proclivis*	+	+				+	+	+		+		
25	Christiania aff. C. scolia	+											
26	Christiania sp. 1	+	+					+					
27	Christiania sp. 2	+								+	+		
28	Craspedelia roomusoksi	+					+						
29	Doughlatomena splendens*		+					+					
30	Dulankarella magna	+	+					+					
31	Dulankarella cf. D. magna		+							+		+	
32	Dzhebaglina plicata	+						+					
33	Eoplectodonta? sp.				+								
34	Eostropheodonta bublitschenki				+								+
35	Eostropheodonta sp.				+								+

Table 1. (continued)

	Global stages				Lithostratigraphical units							
	Katian											
	Ka1	Ka2-Ka3	Ka4	Hirnantian	Anderken Fm.	Kopkurgan Fm.	Otar Mb	Degeres Mb.	Akkol Mb.	Sartan-Manai Lm.	Ulkuntas Fm.	Zhalair Fm.
	2	3	4	5	6	7	8	9	10	11	12	13
36 Foliomena prisca	+				+							
37 Glyptambonites aff. musculosus		+						+	+			
38 Glyptomenoides? sp.		+				+		+				
39 Gunningblandella sp.*		+						+				
40 Holtedahlina orientalis		+					+	+				
41 Karomena squalida	+											
42 Kassinella simorini	+				+							
43 Kassinella sp.			+								+	
44 Leangella paletsae		+						+				
45 Leangella sp.*		+				+						
46 Leptaena (L.) rugosa				+								+
47 Leptaena (L.) aff. L. rugosa			+								+	
48 Leptaena (L.) trifidum				+								+
49 Leptaena (Ygdrasilomena) sp. 2	+									+		
50 Mabella multicostata	+							+				
51 Metambonites subcarinatus	+									+		
52 Metambonites sp.			+									
53 Nikitinamena bicostata	+		+					+				
54 Olgambonites insolita	+				+							
55 Paromalomena polonica				+								
56 Platymena tersa	+	+					+	+				
57 Rafinesquina ultrix				+								+
58 Rafinesquina aff. R. urbicola				+				+				+
59 Rhipidomena sp.		+				+		+				
60 Shlyginia extraordinaria*	+	+						+		+		
61 Shlyginia perplexa	+									+		
62 Sortanella quinquecostata	+	+						+		+		
63 Sowerbyella (S.) akdombakensis	+	+					+	+		+		
64 Sowerbyella (S.) ampla*	+		+			+		+	+		+	
65 Sowerbyella (Rugosowerbyella) cf. R. ambigua								+				
66 Sowerbyella (R.) sp.									+			
67 Strophomena orthonurensis		+					+	+				
68 Strophomena cf. S. orthonurensis	+											
69 Zhilgyzambonites extenuata	+				+							
Order ORTHOTETIDA												
70 Cliftonia sp. 1				+							+	
71 Cliftonia sp. 2			+									+
72 Coolinia illiensis				+							+	
73 Coolinia sp.			+									+

Table 1. (continued)

	Global stages				Lithostratigraphical units								
	Katian											
	Ka1	Ka2–Ka3	Ka4	Hirnantian	Anderken Fm.	Kopkurgan Fm.	Otar Mb	Degeres Mb.	Akkol Mb.	Sartan-Manai Lm.	Ulkuntas Fm.	Zhalair Fm.
1	2	3	4	5	6	7	8	9	10	11	12	13
74 *Gacella* sp.*		+										
75 *Grammoplecia globosa*		+				+						
76 *Grammoplecia* aff. *G. globosa*						+						
77 *Grammoplecia subcraegensis*	+	+					+	+				
78 *Grammoplecia* sp.*		+						+				
79 *Ogmoplecia nesca*		+							+			
80 *Placotriplesia* sp.		+										
81 *Streptis altosinuata*			+								+	
82 *Triplesia sortanensis*	+									+		
83 *Triplesia* sp.			+								+	
Order ORTHIDA												
84 *Austinella?* sp.	+									+		
85 *Bokotorthis kasachstanica*		+				+	+		+			
86 *Dalmanella cicatrica*				+								+
87 *Dalmanella testudinaria*			+								+	
88 *Dalmanella* aff. *D. pectenoides*				+								+
89 *Dinorthis kassini*	+						+					
90 *Dolerorthis* aff. *D. hubeiensis*	+								+	+		
91 *Dolerorthis* sp.		+										+
92 *Epitomyonia* sp.		+						+			+	
93 *Giraldibella* aff. *G. bella*			+									
94 *Hirnantia sagittifera*				+								
95 *Lictorthis licta*		+						+				
96 *Lictorthis* sp.*		+										
97 *Onniella* sp.*		+						+				
98 *Phaceloorthis? corrugata**		+				+						
99 *Plaesiomys fidelis*		+				+	+					
100 *Phragmorthis* sp.		+				+	+					
101 *Plaesiomys fidelis*		+										
102 *Ptychopleurella ramifera*		+								+		
103 *Ptychopleurella?* sp.		+							+			
104 *Weberorthis brevis*	+	+					+	+				
Order PENTAMERIDA												
105 *Holorhynchus giganteus*			+								+	
106 *Parastrophina angulosa nucula*	+									+		
107 *Parastrophina portentosa*	+									+		
108 *Parastrophina tersa*	+									+		
109 *Parastrophina* sp.	+		+									
110 *Proconchidium tchuilense*			+								+	
111 *Tcherskidium? ulkuntasense*											+	
Order RHYNCHONELLIDA												

Table 1. (continued)

	Global stages				Lithostratigraphical units							
	Katian											
	Ka1	Ka2–Ka3	Ka4	Hirnantian	Anderken Fm.	Kopkurgan Fm.	Otar Mb	Degeres Mb.	Akkol Mb.	Sartan-Manai Lm.	Ulkuntas Fm.	Zhalair Fm.
1	2	3	4	5	6	7	8	9	10	11	12	13
112 *Lydirhyncha tarimensis*	+	+				+	+	+	+			
113 *Lydirhyncha* sp.			+								+	
114 *Rostricellula sarysuica*		+								+		
Order ATRYPIDA												
115 *Eospirigerina plana*			+								+	
116 *Kellerella ditissima*	+									+		
117 *Nikolaispira rasilis*	+									+		
118 *Qilianotryma suspectum*		+						+				
119 *Rongatrypa rudis*	+						+					
120 *Schachriomonia parva*		+							+			
121 *Sulcatospira asiatica*	+						+					
122 *Sulcatospira? dominanta*			+								+	
Order ATHYRIDIDA												
123 *Hindella* sp.			+								+	
124 *Iliella minima*			+								+	

*Taxa described or discussed in the present paper.